Functional Foods for Disease Prevention I

Fruits, Vegetables, and Teas

ACS SYMPOSIUM SERIES **701**

Functional Foods for Disease Prevention I

Fruits, Vegetables, and Teas

Takayuki Shibamoto, EDITOR
University of California at Davis

Junji Terao, EDITOR
University of Tokushima

Toshihiko Osawa, EDITOR
Nagoya University

Developed from a symposium sponsored by the Division
of Agricultural and Food Chemistry at the 213th National Meeting
of the American Chemical Society,
San Francisco, California,
April 13–17, 1997

American Chemical Society, Washington, DC

Chemistry Library

Library of Congress Cataloging-in-Publication Data

Functional foods for disease prevention / Takayuki Shibamoto, Junji Tarao,
Toshihiko Osawa, editors.

    p.   cm.—(ACS symposium series, ISSN 0097–6156; 701–702)

    "Developed from a symposium sponsored by the Division of Agricultural and
Food Chemistry at the 213th National Meeting of the American Chemical
Society, San Francisco, California, April 13–17, 1997."

    Includes bibliographical references and indexes.

    Contents: I. Fruits, vegetables, and teas — II. Medicinal plants and other
foods.

    ISBN 0–8412–3572–4 (v. 1). — ISBN 0-8412–3573–2 (v. 2)

    1. Medicinal plants—Congresses. 2. Functional foods—Congresses.

    I. Shibamoto, Takayuki. II. Terao, Junji. III. Osawa, Toshihiko. IV. American
Chemical Society. Division of Agricultural and Food Chemistry. V. American
Chemical Society. Meeting (213th : 1997 : San Francisco, Calif.) VI. Series.

RS164.F87   1998
615′.32—dc21   98–6978
                                               CIP

The paper used in this publication meets the minimum requirements of American National Standard for
Information Sciences—Permanence of Paper for Printed Library Materials. ANSI Z39.48–1984.

PRINTED IN THE UNITED STATES OF AMERICA

Foreword

THE ACS SYMPOSIUM SERIES was first published in 1974 to provide a mechanism for publishing symposia quickly in book form. The purpose of the series is to publish timely, comprehensive books developed from ACS sponsored symposia based on current scientific research. Occasionally, books are developed from symposia sponsored by other organizations when the topic is of keen interest to the chemistry audience.

Before agreeing to publish a book, the proposed table of contents is reviewed for appropriate and comprehensive coverage and for interest to the audience. Some papers may be excluded in order to better focus the book; others may be added to provide comprehensiveness. When appropriate, overview or introductory chapters are added. Drafts of chapters are peer-reviewed prior to final acceptance or rejection, and manuscripts are prepared in camera-ready format.

As a rule, only original research papers and original review papers are included in the volumes. Verbatim reproductions of previously published papers are not accepted.

ACS BOOKS DEPARTMENT

Contents

Tea and Related Compounds

INDEXES

Preface

"FUNCTIONAL FOODS" are called by many different names, including designer foods, pharmafoods, nutraceuticals, medical foods, and a host of other names, depending on one's background and perspective. Recently, food components that possess such biological characteristics, such as anticarcinogenicity, antimutagenicity, antioxidative activity, and antiaging activity, have received much attention from food and nutrition scientists as a third functional component of foods, after nutrients and flavor compounds. This symposium focuses on the latest scientific research and the impact of this research on policy and regulation of functional foods. A major objective of the symposium was to provide a forum for interaction among food chemists, nutritionists, medical doctors, students, policy makers, and interested personnel from industries.

The two volumes of Functional Foods: Overview and Disease Prevention cover the most recent research results and state-of-the art research methodology in the field. In addition, current perspectives on functional foods and regulatory issues are also presented. The contributors are experts in the area of functional foods and were selected from many countries, including the United States, Canada, the Netherlands, Germany, Finland, Israel, Japan, Korea, Thailand, India, and Taiwan.

This book provides valuable information and useful research tools for diverse areas of scientists, including biologists, biochemists, chemists, medical doctors, pharmacologists, nutritionists, and food scientists, from academic institutions, governmental institutions, and private industries.

This volume contains perspectives and the role of functional foods in various human disease prevention. Biological activities of different natural plants such as fruits, vegetables, teas, and their related compounds are presented.

Acknowledgments

We appreciate contributors of this book and participants of the Symposium very much. Without their effort and support, the Symposium would not be successful. We thank Tomoko Shibamoto, Sangeeta Patel, and Takashi Miyake for their assistance in organizing and proceeding the Symposium. We are indebted to Hiromoto Ochi who donated a fund to award 10 outstanding papers.

We acknowledge the financial support of the following sponsors: AIM International Inc.; Amway Japan Ltd.; Avron Resources Inc.; Fuji Oil Co., Ltd.; Green Foods Corporation; Kikkoman Corporation; Mercian Corporation; Mitsui Norin Co., Ltd.; Morinage & Co., Ltd.; Nikken Fine Chemical Co., Ltd.; Nikken Foods Co., Ltd.; Taiho Pharmaceutical Co., Ltd.; The Calpis Food Industry Co., Ltd.; The Rehnborg Center for Nutrition & Wellness and Nutrilite, Amway Corp.; UOP; Yamanouchi Pharmaceutical Co., Ltd.; and The Division of Agricultural and Food Chemistry of the American Chemical Society.

TAKAYUKI SHIBAMOTO
University of California at Davis
Department of Environmental Toxicology
Davis, CA 95616

JUNJI TERAO
Department of Nutrition, School of Medicine
University of Tokushima
Tokushima 305, Japan

TOSHIHIKO OSAWA
Department of Applied Biological Sciences
Faculty of Agriculture
Nagoya University
Chikusa, Nagoya 770-0042, Japan

Contents (Volume II)

PERSPECTIVE AND OVERVIEW

Chapter 1

The Prevention of Cancer

Bruce N. Ames[1] and Lois Swirsky Gold[1,2]

[1]Barker Hall and [2]Lawrence Berkeley National Laboratory, University
of California, Berkeley, CA 94720

1. The major causes of cancer are:
   a) Smoking: About a third of U.S. cancer (90% of lung cancer);
   b) Dietary imbalance, e.g., lack of dietary fruits & vegetables:
The quarter of the population eating the least fruits & vegetables has
double the cancer rate for most types of cancer compared to the quarter
eating the most;
   c) Chronic infections: mostly in developing countries;
   d) Hormonal factors: primarily influenced by life style.
2. There is no epidemic of cancer, except for lung cancer due to
smoking. Overall cancer mortality rates have declined 16% since 1950
(excluding lung cancer).
3. Recent research on animal cancer tests indicates that:
   a) Rodent carcinogens are not rare. Half of all chemicals tested in
standard high-dose animal cancer tests, whether occurring naturally or
produced synthetically, are carcinogens under the test conditions;
   b) There are high-dose effects in rodent cancer tests that are not
relevant to low-dose human exposures and that contribute to the high
proportion of chemicals that test positive;
   c) The focus of regulatory policy is on synthetic chemicals,
although 99.9% of the chemicals humans ingest are natural. Over
1000 chemicals have been described in coffee: 28 have been tested
and 19 are rodent carcinogens. Plants in the human diet contain
thousands of natural pesticides which protect them from insects and
other predators: 63 have been tested and 35 are rodent carcinogens.
4. There is no convincing evidence that synthetic chemical pollutants
are important for human cancer.

Cancer Trends

Cancer death rates overall in the U.S. (excluding lung cancer due to smoking) have
declined 16% since 1950 *(1)*. The types of cancer deaths that have decreased since
1950 are primarily stomach, cervical, uterine, and colorectal. The types that have
increased are primarily lung cancer (90% is due to smoking, as are 35 % of all cancer

deaths in the U.S.), melanoma (probably due to sunburns), and non-Hodgkin's lymphoma. If lung cancer is included, mortality rates have increased over time, but recently have declined in men due to the effects of decreased smoking *(1)*. The rise in incidence rates in older age groups for some cancers, e.g., prostate, can be explained by known factors such as improved screening. "The reason for not focusing on the reported incidence of cancer is that the scope and precision of diagnostic information, practices in screening and early detection, and criteria for reporting cancer have changed so much over time that trends in incidence are not reliable" *(2)*. (See also *(3)* and *(4)*).

Cancer is one of the degenerative diseases of old age and increases exponentially with age in both rodents and humans. External factors, however, can markedly increase cancer rates (e.g., cigarette smoking in humans) or decrease them (e.g., caloric restriction in rodents). Life expectancy has continued to rise since 1950. Thus the increases in the crude number of observed cancer deaths (not adjusted for age) reflect the aging of the population and the delayed effects of earlier increases in smoking *(3,4)*.

Important Causes of Human Cancer

Epidemiological studies have identified the factors that are likely to have a major effect on lowering rates of cancer: reducing smoking, improving diet (e.g., increased consumption of fruits and vegetables), and controlling infections *(5)*. We *(5)* estimate that diet accounts for about one-third of cancer risk in agreement with the earlier estimates of Doll and Peto *(3)*, and we discuss diet in the next section. Other factors are lifestyle influences on hormones, avoidance of intense sun exposure, increased physical activity, reduced consumption of alcohol, and occupational exposures.

Since cancer is due in part to normal aging, to the extent that the major external risk factors for cancer are diminished, (smoking, unbalanced diet, chronic infection, and hormonal factors) cancer will occur at a later age, and the proportion of cancer caused by normal metabolic processes will increase. Aging and its degenerative diseases appear to be due in good part to the accumulation of oxidative damage to DNA and other macromolecules *(6)*. By-products of normal metabolism -- superoxide, hydrogen peroxide, and hydroxyl radical -- are the same oxidative mutagens produced by radiation. An old rat has about 66,000 oxidative DNA lesions per cell *(7)*. DNA is oxidized in normal metabolism because antioxidant defenses, though numerous, are not perfect. Antioxidant defenses against oxidative damage include Vitamins C and E and probably carotenoids, most of which come from dietary fruits and vegetables.

Smoking contributes to about 35% of U.S. cancer, about one-quarter of heart disease, and about 400,000 premature deaths per year in the United States *(8)*. Tobacco is a known cause of cancer of the lung, bladder, mouth, pharynx, pancreas, stomach, larynx, esophagus and possibly colon. Tobacco causes even more deaths by diseases other than cancer. Smoke contains a wide variety of mutagens and rodent carcinogens. Smoking is also a severe oxidative stress and causes inflammation in the lung. The oxidants in cigarette smoke--mainly nitrogen oxides--deplete the body's antioxidants. Thus, smokers must ingest two to three times more vitamin C than non-smokers to achieve the same level in blood, but they rarely do. Inadequate concentration of Vitamin C in plasma is more common among the poor and smokers *(6)*.

Men with inadequate diets or who smoke may damage both their somatic DNA and the DNA of their sperm. When the dietary Vitamin C is insufficient to keep seminal fluid Vitamin C at an adequate level, the oxidative lesions in sperm DNA are increased 250% *(9-11)*. Smokers also produce more aneuploid sperm than non-smokers *(12)*. Paternal smokers, therefore, may plausibly increase the risk of birth

4

defects and childhood cancer in offspring *(9,10)*. New epidemiological evidence indicates that childhood cancers are increased in offspring of male smokers, e.g., acute lymphocytic leukemia, lymphoma, and brain tumors, are increased 3-4 times *(13)*.

Chronic inflammation from chronic infection results in release of oxidative mutagens from phagocytic cells and is a major contributor to cancer *(5,14)*. White cells and other phagocytic cells of the immune system combat bacteria, parasites, and virus-infected cells by destroying them with potent, mutagenic oxidizing agents. The oxidants protect humans from immediate death from infection, but they also cause oxidative damage to DNA, chronic cell killing with compensatory cell division *(15)* and thus contribute to the carcinogenic process. Antioxidants appear to inhibit some of the pathology of chronic inflammation. Chronic infections cause about 21% of new cancer cases in developing countries and 9% in developed countries *(16)*

Endogenous reproductive hormones play a large role in cancer, including cancer of the breast, prostate, ovary and endometrium *(17,18)*, contributing to as much as 20% of all cancer. Many lifestyle factors such as reproductive history, lack of exercise, obesity, and alcohol influence hormone levels and therefore risk *(5,17-20)*.

Genetic factors also play a significant role and interact with lifestyle and other risk factors. Biomedical research is uncovering important genetic variation in humans.

Occupational exposure to carcinogens can cause cancer, though how much has been a controversial issue; a few percent seems a reasonable estimate *(5)*. The main contributor was asbestos in smokers. Workplace exposures can be high in comparison with other chemical exposures in food, air, or water. Past occupational exposures have sometimes been high and therefore comparatively little quantitative extrapolation may be required for risk assessment from high-dose rodent tests to high-dose occupational exposures. Since occupational cancer is concentrated among small groups exposed at high levels, there is an opportunity to control or eliminate risks once they are identified.

Although some epidemiologic studies find an association between cancer and low levels of industrial pollutants, the associations are usually weak, the results are usually conflicting, and the studies do not correct for potentially large confounding factors like diet. Moreover, the exposures to synthetic pollutants are small and the low concentrations do not seem plausible as a causal factor when compared to the background of natural chemicals that are rodent carcinogens *(21)*. Even assuming that the EPA's worst-case risk estimates for synthetic pollutants are true risks, the proportion of cancer that EPA could prevent by regulation would be tiny *(22)*.

Preventing Diet-Related Cancer

High consumption of fruits and vegetables is associated with a lowered risk of degenerative diseases including cancer, cardiovascular disease, cataracts, and brain dysfunction *(6)*. More than 200 studies in the epidemiological literature have been reviewed that show, with great consistency, an association between low consumption of fruits and vegetables and cancer incidence *(23-25)* (Table 1). The quarter of the population with the lowest dietary intake of fruits and vegetables compared to the quarter with the highest intake has roughly twice the cancer rate for most types of cancer (lung, larynx, oral cavity, esophagus, stomach, colorectal, bladder, pancreas, cervix, and ovary). Eighty percent of American children and adolescents and 68% of adults *(26,27)* did not meet the intake recommended by the NCI and the National Research Council: 5 servings of fruits and vegetables per day. Publicity about hundreds of minor hypothetical risks can cause loss of perspective on what is important for disease prevention: half the public does not know that fruit and vegetable consumption is a major protection against cancer *(28)*.

Table I. Review of epidemiological studies on cancer showing protection by consumption of fruits and vegetables

Cancer site	Fraction of studies showing significant cancer protection	Relative risk (median) Low vs. high quartile of consumption
Epithelial		
Lung	24/25	2.2
Oral	9/9	2.0
Larynx	4/4	2.3
Esophagus	15/16	2.0
Stomach	17/19	2.5
Pancreas	9/11	2.8
Cervix	7/8	2.0
Bladder	3/5	2.1
Colorectal	20/35	1.9
Miscellaneous	6/8	---
Hormone-dependent		
Breast	8/14	1.3
Ovary/endometrium	3/4	1.8
Prostate	4/14	1.3
Total	129/172	

SOURCE: Adapted from ref. *(23)*.

Micronutrients in Fruits and Vegetables are Anticarcinogens.
Antioxidants in fruits and vegetables may account for some of their beneficial effect, as discussed above. However, the effects of dietary antioxidants are difficult to disentangle by epidemiological studies from other important vitamins and ingredients in fruits and vegetables *(23,24,29,30)*.

Folate deficiency, one of the most common vitamin deficiencies, causes chromosome breaks in human genes *(31)*. Approximately 10% of the US population *(32)* is deficient at the level causing chromosome breaks. In two small studies of low income (mainly African-American) elderly *(33)* and adolescents *(34)* nearly half had folate levels that low. The mechanism is deficient methylation of uracil to thymine, and subsequent incorporation of uracil into human DNA (4 million/cell) *(31)*. During repair of uracil in DNA, transient nicks are formed; two opposing nicks causes a chromosome break. Both high DNA uracil levels and chromosome breaks in humans are reversed by folate administration *(31)*. Chromosome breaks could contribute to the increased risk of cancer and cognitive defects associated with folate deficiency in humans *(31)*. Folate deficiency also damages human sperm *(35)*, causes neural tube defects in the fetus, and about 10% of the risk of heart disease in the U.S. *(31)*. Diets deficient in fruits and vegetables are commonly low in folate, antioxidants, (e.g., Vitamin C) and many other micronutrients, and result in significant amounts of DNA damage and higher cancer rates *(5,23,36)*.

Other micronutrients, whose main dietary sources are other than fruits and vegetables, are likely to play a significant role in the prevention and repair of DNA damage, and thus are important to the maintenance of long term health. Deficiency of

Vitamin B_{12} causes a functional folate deficiency, accumulation of homocysteine (a risk factor for heart disease) *(37)*, and misincorporation of uracil into DNA *(38)*. Strict vegetarians are at increased risk of developing a Vitamin B_{12} deficiency *(37)*. Niacin contributes to the repair of DNA strand breaks by maintaining nicotinamide adenine dinucleotide levels for the poly ADP-ribose protective response to DNA damage *(39)*. As a result, dietary insufficiencies of niacin (15% of some populations are deficient *(40)*), folate, and antioxidants may act synergistically to adversely affect DNA synthesis and repair.

Optimizing micronutrient intake can have a major impact on health at low cost. Increasing research in this area and efforts to improve micronutrient intake and balanced diet should be a high priority for public policy. Fruits and vegetables are of major importance for reducing cancer: if they become more expensive by reducing use of synthetic pesticides, cancer is likely to increase. People with low incomes eat fewer fruits and vegetables and spend a higher percentage of their income on food.

Calories or Protein Restriction and Cancer Prevention. In rodents a calorie-restricted diet, compared to *ad libitum* feeding, markedly decreases tumor incidence and increases lifespan, but decreases reproduction *(41,42)*. Protein restriction, though less well-studied, appears to have similar effects *(43)*. Darwinian fitness in animals appears to be increased by hormonal changes which delay reproductive function during periods of low food availability because the saved resources are invested in maintenance of the body until food resources are available for successful reproduction *(44,45)*. Lower mitotic rates are observed in a variety of tissues in calorie-restricted compared to *ad libitum* fed rodents *(46,47)*, which is likely to contribute to the decrease in tumor incidence *(48)*. Though epidemiological evidence on restriction in humans is sparse, the possible importance of growth restriction in human cancer is supported by epidemiologic studies indicating higher rates of breast and other cancers among taller persons *(49)*; e.g., Japanese women are now taller, menstruate earlier, and have increased breast cancer rates. Also, many of the variations in breast cancer rates among countries, and trends over time within countries, are compatible with changes in growth rates and attained adult height *(50)*. Obesity in post menopausal women is a risk factor for breast cancer *(20,49)*.

Are Human Exposures to Pollutants or Pesticide Residues that are Rodent Carcinogens Likely to be Important for Human Cancer?

There is an enormous background of human exposures to naturally-occurring chemicals, and half of natural (as well as synthetic) chemicals tested are rodent carcinogens. 99.9% of the chemicals humans ingest are natural. The amounts of synthetic pesticide residues in plant foods are insignificant compared to the amount of natural pesticides produced by plants themselves *(51,52)*. Of all dietary pesticides that humans eat, 99.99% are natural: they are chemicals produced by plants to defend themselves against fungi, insects, and other animal predators *(51,52)*. Each plant produces a different array of such chemicals. On average Americans ingest roughly 5,000 to 10,000 different natural pesticides and their breakdown products. Americans eat about 1,500 mg of natural pesticides per person per day, which is about 10,000 times more than they consume of synthetic pesticide residues.

Even though only a small proportion of natural pesticides has been tested for carcinogenicity, half of those tested (35/64) are rodent carcinogens, and naturally occurring pesticides that are rodent carcinogens are ubiquitous in fruits, vegetables, herbs, and spices *(53)* (Table II).

Cooking foods produces about 2,000 mg per person per day of burnt material that contains many rodent carcinogens and many mutagens. By contrast, the residues of 200 synthetic chemicals measured by FDA, including the synthetic pesticides thought to be of greatest importance, average only about 0.09 mg per person per day *(51,53)*. The known natural rodent carcinogens in a single cup of coffee are about equal in weight to an entire year's worth of synthetic pesticide residues that are rodent carcinogens, even though only 3% of the natural chemicals in roasted coffee have been tested for carcinogenicity *(21)* This does not mean that coffee is dangerous, but rather that assumptions about high-dose animal cancer tests for assessing human risk at low doses need reexamination. No diet can be free of natural chemicals that are rodent carcinogens *(53)*.

Table II. Carcinogenicity of natural plant pesticides tested in rodents (Fungal toxins are not included)

Carcinogens: N=35	acetaldehyde methylformylhydrazone, allyl isothiocyanate, arecoline.HCl, benzaldehyde, benzyl acetate, caffeic acid, catechol, clivorine, coumarin, crotonaldehyde, cycasin and methylazoxymethanol acetate, 3,4-dihydrocoumarin, estragole, ethyl acrylate, $N2$-γ-glutamyl-p-hydrazinobenzoic acid, hexanal methylformylhydrazine, p-hydrazinobenzoic acid.HCl, hydroquinone, 1-hydroxyanthraquinone, lasiocarpine, d-limonene, 8-methoxypsoralen, N-methyl-N-formylhydrazine, α-methylbenzyl alcohol, 3-methylbutanal methylformylhydrazone, methylhydrazine, monocrotaline, pentanal methylformylhydrazone, petasitenine, quercetin, reserpine, safrole, senkirkine, sesamol, symphytine
Noncarcinogens: N=28	atropine, benzyl alcohol, biphenyl, d-carvone, deserpidine, disodium glycyrrhizinate, emetine.2HCl, ephedrine sulphate, eucalyptol, eugenol, gallic acid, geranyl acetate, β-N-[γ-l(+)-glutamyl]-4-hydroxy-methylphenylhydrazine, glycyrrhetinic acid, p-hydrazinobenzoic acid, isosafrole, kaempferol, dl-menthol, nicotine, norharman, pilocarpine, piperidine, protocatechuic acid, rotenone, rutin sulfate, sodium benzoate, turmeric oleoresin, vinblastine

These rodent carcinogens occur in: absinthe, allspice, anise, apple, apricot, banana, basil, beet, broccoli, Brussels sprouts, cabbage, cantaloupe, caraway, cardamom, carrot, cauliflower, celery, cherries, chili pepper, chocolate milk, cinnamon, cloves, cocoa, coffee, collard greens, comfrey herb tea, corn, coriander, currants, dill, eggplant, endive, fennel, garlic, grapefruit, grapes, guava, honey, honeydew melon, horseradish, kale, lemon, lentils, lettuce, licorice, lime, mace, mango, marjoram, mint, mushrooms, mustard, nutmeg, onion, orange, paprika, parsley, parsnip, peach, pear, peas, black pepper, pineapple, plum, potato, radish, raspberries, rhubarb, rosemary, rutabaga, sage, savory, sesame seeds, soybean, star anise, tarragon, tea, thyme, tomato, turmeric, and turnip.

SOURCE: Adapted from ref. *(53)*.

Why are Half of the Chemicals Tested in High-Dose Animal Cancer Tests Rodent Carcinogens? Approximately half of all chemicals -- whether natural or synthetic -- that have been tested in standard animal cancer tests are rodent carcinogens *(54,55)* (Table III). We have concluded that although there may be some bias in picking more suspicious chemicals such bias is not the major explanation for the high positivity rate *(56,57)*.

Table III. Proportion of chemicals evaluated as carcinogenic.

Chemicals tested in both rats and mice[a]	330/559	(59%)
Naturally-occurring chemicals	73/127	(57%)
Synthetic chemicals	257/432	(59%)
Chemicals tested in rats and/or mice[a]		
Chemicals in Carcinogenic Potency Database	668/1275	(52%)
Natural pesticides	35/63	(56%)
Mold toxins	14/23	(61%)
Chemicals in roasted coffee	19/28	(68%)
Innes negative chemicals retested[a,b]	16/34	(47%)
Physician's Desk Reference (PDR): drugs with reported cancer tests[c]	117/241	(49%)
FDA database of drug submissions[d]	125/282	(44%)

SOURCES: [a] The Carcinogenic Potency Database, adapted from ref. *(55)*.
[b] The 1969 study by Innes *et al.,* adapted from ref. *(64)*, is frequently cited as evidence that the proportion of carcinogens is low, as only 9% of 119 chemicals tested (primarily pesticides) were positive. However, these tests, which were only in mice with few animals per group, lacked the power of modern tests.
[c] Adapted from ref. *(65)*, Davies and Monro
[d] Adapted from ref. *(66)*, Contrera *et al.* 140 drugs are in both the FDA and PDR databases.

In standard cancer tests rodents are given chronic, near-toxic doses, the maximum tolerated dose (MTD). Evidence is accumulating that it may be cell division caused by the high dose itself, rather than the chemical per se, that is increasing the positivity rate. Endogenous DNA damage from normal oxidation is large. Thus, from first principles, the cell division rate must be a factor in converting lesions to mutations and thus cancer *(58)*. Raising the level of either DNA lesions or cell division will increase the probability of cancer. Just as DNA repair protects against lesions, p53 guards the cell cycle and defends against cell division if the lesion level gets too high *(5)*. If the lesion level becomes higher still, p53 can initiate programmed cell death (apoptosis) *(59,60)*. None of these defenses is perfect, however *(5)*. The critical factor is chronic cell division in stem cells, not in cells that are discarded, and is related to the total number of extra cell divisions *(61)*. Cell division is both a major factor in loss of heterozygosity through non-disjunction and other mechanisms *(62,63)* and in expanding clones of mutated cells.

High doses can cause chronic wounding of tissues, cell death, and consequent chronic cell division of neighboring cells, which is a risk factor for cancer *(54)*. Tissues injured by high doses of chemicals have an inflammatory immune response involving activation of recruited and resident macrophages *(67-73)* (e.g., phenobarbital, carbon tetrachloride, TPA). Activated macrophages release mutagenic oxidants (including peroxynitrite, hypochlorite, and H_2O_2), as well as inflammatory and cytotoxic cytokines, growth factors, bioactive lipids (arachidonic acid metabolites), and proteases. This general response to cell injury suggests that chronic cell killing by high dose animal cancer tests will likely incite a similar response, leading to further cell injury, compensatory cell division and therefore increased probability of mutation.

Thus it seems likely that a high proportion of all chemicals, whether synthetic or natural, might be "carcinogens" if run through the standard rodent bioassay at the MTD, but this will be primarily due to the effects of high doses for the non-mutagens, and a synergistic effect of cell division at high doses with DNA damage for the mutagens *(58,63,74)*.

Correlation between Cell Division and Cancer. Many studies on rodent carcinogenicity show a correlation between cell division at the MTD and cancer. Cunningham *et al.* have analyzed 15 chemicals at the MTD, 8 mutagens and 7 non-mutagens, including pairs of mutagenic isomers, one of which is a carcinogen and one of which is not *(75-85)*. They found a perfect correlation between cancer causation and cell division in the target tissue: the 9 chemicals increasing cancer caused cell division in the target tissue and the 6 chemicals not increasing cancer did not. A similar result has been found in the analyses of Mirsalis *(86)*, e.g., both dimethylnitrosamine (DMN) and methyl methane sulfonate (MMS) methylate liver DNA and cause unscheduled DNA synthesis (a result of DNA repair), but DMN causes both cell division and liver tumors, while MMS does neither. A recent study on the mutagenic dose response of the carcinogen ethylnitrosourea concludes that cell division is a key factor in its mutagenesis and carcinogenesis *(87)*. Chloroform at high doses induces liver cancer by chronic cell division *(88)*. Formaldehyde causes cancer at high doses, primarily through increases in cell division *(61)*. PhIP, a mutagenic heterocyclic amine from cooked protein, induces colon tumors in male rats, but not in female rats; the level of DNA adducts in the colonic mucosa was the same in both sexes, however, cell division was increased only in the male, contributing to the formation of premalignant lesions of the colon *(89)*. Therefore, there was no correlation between adduct formation and these premalignant lesions, but there was between cell division and lesions. The importance of cell division for a variety of genotoxic and non-genotoxic agents has been demonstrated *(90)*. Extensive reviews on rodent studies *(58,63,91-94)* document that chronic cell division can induce cancer. There is also a large epidemiological literature reviewed by Preston-Martin, Henderson and colleagues *(95,96)* showing that increased cell division by hormones and other agents can increase human cancer. At the low levels to which humans are usually exposed, such increased cell division does not occur. Therefore, the very low levels of chemicals to which humans are exposed through water pollution or synthetic pesticide residues are likely to pose no or minimal cancer risks.

Risk Assessment. In regulatory policy, the "virtually safe dose" (VSD), corresponding to a maximum, hypothetical cancer risk of one in a million, is estimated from bioassay results using a linear model. To the extent that carcinogenicity in rodent bioassays is due to the effects of high doses for the non-mutagens, and a synergistic effect of cell division at high doses with DNA damage for the mutagens, then this model is inappropriate. As we pointed out in 1990 *(63)*: "The high proportion of carcinogens among chemicals tested at the MTD emphasizes the importance of understanding cancer mechanisms in order to determine the relevance of rodent cancer test results for humans. A list of rodent carcinogens is not enough. The main rule in toxicology is that 'the dose makes the poison': at some level, every chemical becomes toxic, but there are safe levels below that. However, the precedent of radiation, which is both a mutagen and a carcinogen, gave credence to the idea that there could be effects of chemicals even at low doses. A scientific consensus evolved in the 1970s that we should treat carcinogens differently, that we should assume that even low doses might cause cancer, even though we lacked the methods for measuring carcinogenic effects at low levels. This idea evolved because it was expected that (*i*) only a small proportion of chemicals would have carcinogenic potential, (*ii*) testing at a high dose would not produce a carcinogenic effect unique to the high dose, and (*iii*) chemical carcinogenesis would be explained by the mutagenic potential of chemicals. However, it seems time to take account of new information suggesting that all three assumptions are wrong."

Possible Hazards from Synthetic Chemicals Should be Viewed in the Context of Natural Chemicals. Gaining a broad perspective about the vast number of chemicals to which humans are exposed can be helpful when setting research and regulatory priorities (*21,52,97,98*). Rodent bioassays provide little information about mechanisms of carcinogenesis and low-dose risk. The assumption that synthetic chemicals are hazardous has led to a bias in testing, such that synthetic chemicals account for 77% of the 559 chemicals tested chronically in both rats and mice (Table 3). The natural world of chemicals has never been tested systematically. One reasonable strategy is to use a rough index to *compare* and *rank* possible carcinogenic hazards from a wide variety of chemical exposures at levels that humans typically receive, and then to focus on those that rank highest (*21,98,99*). We have ranked 74 human exposures to rodent carcinogens using the HERP index (Human Exposure/Rodent Potency), which indicates what percentage of the rodent potency (Tumorigenic Dose rate for 50% of rodents, TD_{50} in mg/kg/day) a human receives from a given daily lifetime exposure (mg/kg/day). Overall, our analyses have shown that HERP values for some historically high exposures in the workplace and some pharmaceuticals rank high, and that there is an enormous background of naturally occurring rodent carcinogens in typical portions of common foods that cast doubt on the relative importance of low-dose exposures to residues of synthetic chemicals such as pesticides (*21,55,98,100*). A committee of the National Research Council/National Academy of Sciences recently reached similar conclusions about natural vs. synthetic chemicals in the diet, and called for further research on natural chemicals (*101*).

The possible carcinogenic hazards from synthetic pesticides (at average exposures) are minimal compared to the background of nature's pesticides, though neither may be a hazard at the low doses consumed. This analysis also indicates that many ordinary foods would not pass the regulatory criteria used for synthetic chemicals. Our results call for a re-evaluation of the utility of animal cancer tests in protecting the public against minor hypothetical risks.

It is often assumed that because natural chemicals are part of human evolutionary history, whereas synthetic chemicals are recent, the mechanisms that have evolved in animals to cope with the toxicity of natural chemicals will fail to protect against synthetic chemicals. This assumption is flawed for several reasons (*52,54*).

a) Humans have many natural defenses that make us well buffered against normal exposures to toxins (*52*), and these are usually general, rather than tailored for each specific chemical. Thus they work against both natural and synthetic chemicals. Examples of general defenses include the continuous shedding of cells exposed to toxins -- the surface layers of the mouth, esophagus, stomach, intestine, colon, skin, and lungs are discarded every few days; DNA repair enzymes, which repair DNA that has been damaged from many different sources; and detoxification enzymes of the liver and other organs which generally target classes of toxins rather than individual toxins. That defenses are usually general, rather than specific for each chemical, makes good evolutionary sense. The reason that predators of plants evolved general defenses is presumably to be prepared to counter a diverse and ever-changing array of plant toxins in an evolving world; if a herbivore had defenses against only a set of specific toxins, it would be at a great disadvantage in obtaining new food when favored foods became scarce or evolved new toxins.

b) Various natural toxins which have been present throughout vertebrate evolutionary history, nevertheless cause cancer in vertebrates (52,55). Mold toxins, such as aflatoxin, have been shown to cause cancer in rodents and other species including humans (Table 3). Many of the common elements are carcinogenic to humans at high doses (e.g., salts of cadmium, beryllium, nickel, chromium, and arsenic) despite their presence throughout evolution. Furthermore, epidemiological studies from various parts of the world show that certain natural chemicals in food may be carcinogenic risks to humans; for example, the chewing of betel nuts with tobacco has been correlated with oral cancer.

c) Humans have not had time to evolve a "toxic harmony" with all of their dietary plants. The human diet has changed dramatically in the last few thousand years. Indeed, very few of the plants that humans eat today (e.g., coffee, cocoa, tea, potatoes, tomatoes, corn, avocados, mangoes, olives, and kiwi fruit), would have been present in a hunter-gatherer's diet. Natural selection works far too slowly for humans to have evolved specific resistance to the food toxins in these newly introduced plants.

d) DDT is often viewed as the typically dangerous synthetic pesticide because it concentrates in the tissues and persists for years, being slowly released into the bloodstream. DDT, the first synthetic pesticide, eradicated malaria from many parts of the world, including the U.S. It was effective against many vectors of disease such as mosquitoes, tsetse flies, lice, ticks, and fleas. DDT was also lethal to many crop pests, and significantly increased the supply and lowered the cost of food, making nutritious foods more accessible to poor people. It was also of low toxicity to humans. A 1970 National Academy of Sciences report concluded: "In little more than two decades DDT has prevented 500 million deaths due to malaria, that would other wise have been inevitable (102)." There is no convincing epidemiological evidence, nor is there much toxicological plausibility, that the levels normally found in the environment are likely to be a significant contributor to cancer. DDT was unusual with respect to bioconcentration, and because of its chlorine substituents it takes longer to degrade in nature than most chemicals; however, these are properties of relatively few synthetic chemicals. In addition, many thousands of chlorinated chemicals are produced in nature, and natural pesticides also can bioconcentrate if they are fat soluble. Potatoes, for example, naturally contain the fat soluble neurotoxins solanine and chaconine, which can be detected in the bloodstream of all potato eaters. High levels of these potato neurotoxins have been shown to cause malformations in the hamster fetus (103).

e) Since no plot of land is immune to attack by insects, plants need chemical defenses -- either natural or synthetic -- in order to survive pest attack. Thus, there is a trade-off between naturally occurring pesticides and synthetic pesticides. One consequence of disproportionate concern about synthetic pesticide residues is that some plant breeders develop plants to be more insect-resistant by making them higher in natural toxins. A recent case illustrates the potential hazards of this approach to pest control: When a major grower introduced a new variety of highly insect-resistant celery into commerce, people who handled the celery developed rashes when they were subsequently exposed to sunlight. Some detective work found that the pest-resistant celery contained 6,200 parts per billion (ppb) of carcinogenic (and mutagenic) psoralens instead of the 800 ppb present in common celery (52).

Are Pesticides and Other Synthetic Chemicals Disrupting Human Hormones? Hormonal factors are important in cancer (see above). A recent book *(104)*, holds that traces of synthetic chemicals, such as pesticides with weak hormonal activity, may contribute to cancer and reduce sperm counts. This view ignores the facts that the usual diet contains natural chemicals that have estrogenic activity millions of times higher than that due to traces of synthetic estrogenic chemicals *(105,106)* and that lifestyle factors can markedly change the levels of endogenous hormones (see above). The low levels of human exposure to residues of industrial chemicals are toxicologically implausible as a significant cause of cancer or reproductive abnormalities, especially when compared to the natural background *(105-107)*. In addition, even if sperm counts really were declining, which is not all clear *(108)*, there are many more likely causes, such as smoking and diet (see above).

Does Regulation of Low Hypothetical Risks Advance Public Health?

The world is not risk-free, and resources are limited; therefore, society must set priorities based on which risks are most important in order to save the most lives. The EPA reports that its regulations cost society $140 billion per year. It has been argued that overall these regulations harm public health *(109-112)*, because "wealthier is not only healthier but highly risk reducing." One estimate indicates "that for every 1% increase in income, mortality is reduced by 0.05%" *(110,113)*. In addition, the median toxin control program costs 58 times more per life-year saved than the median injury prevention program and 146 times more than the median medical program *(114)*. It has been estimated that the U.S. could prevent 60,000 deaths a year by redirecting resources to more cost effective programs *(115)*. The discrepancy is likely to be greater because cancer risk estimates used for toxin control programs are worst-case, hypothetical estimates, and the true risks at low dose are often likely to be zero *(21,54,55)* (see above).

Regulatory efforts to reduce low-level human exposures to synthetic chemicals are expensive because they aim to eliminate minuscule concentrations that now can be measured with improved techniques. These efforts are distraction from the major task of improving public health through increasing scientific understanding about how to prevent cancer (e.g., the role of diet), increasing public understanding of how lifestyle influences health, and improving our ability to help individuals alter lifestyle.

Rules on air and water pollution are necessary (e.g., it was a public health advance to phase lead out of gasoline), and clearly, cancer prevention is not the only reason for regulations. As we pointed out in 1990 *(116)*: "What is chiefly needed is to take seriously the control of the major hazards that have been reliably identified, without diverting attention from these major causes by a succession of highly publicized scares about factors that may well be of little or no importance as causes of human diseases."

Acknowledgments

This paper has been adapted in part from references *(6, 53, 54, and 116)*. This work was supported by the National Cancer Institute Outstanding Investigator Grant CA39910 to B.N.A., the Office of Energy Research, Office of Health and Environmental Research of the U.S. Department of Energy under Contract DE-AC03-76SF00098 to L.S.G., and the National Institute of Environmental Health Sciences Center Grant ESO1896.

Literature Cited

(1)   L. A. G. Ries, C. L. Kosary, B. F. Hankey, B. A. Miller, A. Harras, and B. K. Edwards. *SEER Cancer Statistics Review, 1973-1994*, National Cancer Institute: Bethesda, MD, 1997.

(2)   I. Bailar, J C, and H. L. Gornik, *N. Engl. J. Med.*, **1997**,*336*, 1569-1574.

(3)   R. Doll, and R. Peto, *J. Natl. Cancer Inst.*, **1981**,*66*, 1191-1308.

(4)   S. S. Devesa, W. J. Blot, B. J. Stone, B. A. Miller, R. E. Tarone, and F. J. Fraumeni, Jr., *J. Natl. Cancer Inst.*, **1995**,*87*, 175-182.

(5)   B. N. Ames, L. S. Gold, and W. C. Willett, *Proc. Natl. Acad. Sci. USA*, **1995**,*92*, 5258-5265.

(6)   B. N. Ames, M. K. Shigenaga, and T. M. Hagen, *Proc. Natl. Acad. Sci. USA*, **1993**,*90*, 7915-7922.

(7)   H. J. Helbock, K. B. Beckman, P. Walter, M. K. Shigenaga, A. A. Woodall, H. C. Yeo, and B. N. Ames, *PNAS*, **1997**,in press.

(8)   R. Peto, A. D. Lopez, J. Boreham, M. Thun, and C. Heath, Jr. *Mortality from Smoking in Developed Countries 1950-2000*, Oxford University Press: Oxford, England, 1994.

(9)   C. G. Fraga, P. A. Motchnik, M. K. Shigenaga, H. J. Helbock, R. A. Jacob, and B. N. Ames, *Proc. Natl. Acad. Sci. USA*, **1991**,*88*, 11003-11006.

(10)   A. A. Woodall, and B. N. Ames. In *Vitamin C in Health and Disease*; Packer, L., ed.; Marcel Dekker, Inc.: New York, 1997; pp 193-203.

(11)   C. G. Fraga, P. A. Motchnik, A. J. Wyrobek, D. M. Rempel, and B. N. Ames, *Mutat. Res.*, **1996**,*351*, 199-203.

(12)   A. J. Wyrobek, J. Rubes, M. Cassel, D. Moore, S. Perrault, V. Slott, D. Evenson, Z. Zudova, L. Borkovec, S. Selevan, and X. Lowe, *Am. J. Hum. Genet.*, **1995**,*57*, 737.

(13)   B.-T. Ji, X.-O. Shu, M. S. Linet, W. Zheng, S. Wacholder, Y.-T. Gao, D.-M. Ying, and F. Jin, *J. Natl. Cancer Inst.*, **1997**,*89*, 238-244.

(14)   S. Christen, T. M. Hagen, M. K. Shigenaga, and B. N. Ames. In *Microbes and Malignancy: Infection as a Cause of Cancer*; Parsonnet, J., Horning, S., eds.; Oxford University Press: Oxford, 1997; pp in press.

(15)   E. Shacter, E. J. Beecham, J. M. Covey, K. W. Kohn, and M. Potter, *Carcinogenesis*, **1988**,*9*, 2297-2304.

(16)   P. Pisani, D. M. Parkin, N. Muñoz, and J. Ferlay, *Cancer Epidemiol. Biomarkers & Prev*, **1997**,*6*, 387-400.

(17)   B. E. Henderson, R. K. Ross, and M. C. Pike, *Science*, **1991**,*254*, 1131-1138.

(18)   H. S. Feigelson, and B. E. Henderson, *Carcinogenesis*, **1996**,*17*, 2279-2284.

(19)   S.-O. Andersson, A. Wolk, R. Bergstrom, H.-O. Adami, G. Engholm, A. Englund, and O. Nyren, *J. Natl. Canc. Inst.*, **1997**,*89*, 385-389.

(20)   N. Potischman, C. A. Swanson, P. Siiteri, and R. N. Hoover, *J. Natl. Canc. Inst.*, **1997**,*89*, 397-398.

(21)   L. S. Gold, T. H. Slone, B. R. Stern, N. B. Manley, and B. N. Ames, *Science*, **1992**,*258*, 261-265.

(22)   M. Gough, *Risk Anal.*, **1990**,*10*, 1-6.

(23)   G. Block, B. Patterson, and A. Subar, *Nutr. and Cancer*, **1992**,*18*, 1-29.

(24)   K. A. Steinmetz, and J. D. Potter, *J. Am. Diet Assoc.*, **1996**,*96*, 1027-1039.

(25)   M. J. Hill, A. Giacosa, and C. P. J. Caygill. *Epidemiology of Diet and Cancer*, Ellis Horwood Limited: West Sussex, England, 1994.

(26)   S. M. Krebs-Smith, A. Cook, A. F. Subar, L. Cleveland, J. Friday, and L. L. Kahle, *Arch. Pediatr. Adolesc. Med.*, **1996**,*150*, 81-86.

(27)   S. M. Krebs-Smith, A. Cook, A. F. Subar, L. Cleveland, and J. Friday, *Am. J. Public Health*, **1995**,*85*, 1623-1629.

(28)   A National Cancer Institute Graphic, *J. Natl. Cancer Inst.*, **1996**,*88*, 1314.

(29)   K. A. Steinmetz, and J. D. Potter, *Cancer Causes Control*, **1991**,*2*, 325-357.

(30)   G. Block, *Nutr. Rev.*, **1992**,*50*, 207-213.

(31)   B. C. Blount, M. M. Mack, C. Wehr, J. MacGregor, R. Hiatt, G. Wang, S. N. Wickramasinghe, R. B. Everson, and B. N. Ames, *Proc. Natl. Acad. Sci. USA*, **1997**,*94*, 3290-3295.

(32)   F. R. Senti, and S. M. Pilch, *J. Nutr.*, **1985**,*115*, 1398-402.

(33)   L. B. Bailey, P. A. Wagner, G. J. Christakis, P. E. Araujo, H. Appledorf, C. G. Davis, J. Masteryanni, and J. S. Dinning, *Am. J. Clin. Nutr.*, **1979**,*32*, 2346-2353.

(34)   L. B. Bailey, P. A. Wagner, G. J. Christakis, C. G. Davis, H. Appledorf, P. E. Araujo, E. Dorsey, and J. S. Dinning, *Am. J. Clin. Nutr.*, **1982**,*35*, 1023-1032.

(35)   L. Wallock, A. Woodall, R. Jacob, and B. Ames. Nutritional status and positive relation of plasma folate to fertility indices in nonsmoking men. FASEB Annual Meeting, Experimental Biology 97 1997, New Orleans, LA: p. A184 (Abstracts).

(36)   A. F. Subar, G. Block, and L. D. James, *Am. J. Clin. Nutr.*, **1989**,*50*, 508-16.

14

(37) V. Herbert. In *Present Knowledge in Nutrition*; Ziegler, E.E., Filer, L.J., eds.; ILSI Press: Washington, D.C., 1996; pp 191-205.

(38) S. N. Wickramasinghe, and S. Fida, *Blood*, **1994**,*83*, 1656-61.

(39) J. Z. Zhang, S. M. Henning, and M. E. Swendseid, *J. Nutr.*, **1993**,*123*, 1349-55.

(40) E. L. Jacobson, *J. Am. Coll. Nutr.*, **1993**,*12*, 412-6.

(41) F. J. C. Roe, *Proc. Nutr. Soc.*, **1981**,*40*, 57-65.

(42) R. K. Boutwell, and M. W. Pariza, *Am. J. Clin. Nutr.*, **1987**,*45 (Suppl)*, 151-156.

(43) L. D. Youngman, J.-Y. K. Park, and B. N. Ames, *Proc. Natl. Acad. Sci. USA*, **1992**,*89*, 9112-9116.

(44) A. M. Holehan, and B. J. Merry, *Mech. Ageing Dev.*, **1985**,*32*, 63-76.

(45) R. Holliday, *Bioessays*, **1989**,*10*, 125-7.

(46) T. D. Heller, P. R. Holt, and A. Richardson, *Gastroenterology*, **1990**,*98*, 387-391.

(47) E. Lok, F. W. Scott, R. Mongeau, E. A. Nera, S. Malcolm, and D. B. Clayson, *Cancer Lett.*, **1990**,*51*, 67-73.

(48) B. Grasl-Kraupp, Bursch, W., Ruttkay-Nedecky, B., Wagner, A., Lauer, B. & Schulte-Hermann, R., *Proc. Natl. Acad. Sci. USA*, **1994**,*91*, 9995-9999.

(49) D. J. Hunter, and W. C. Willett, *Epidemiol. Rev.*, **1993**,*15*, 110-132.

(50) W. C. Willett, and M. J. Stampfer, *Cancer Causes Control*, **1990**,*1*, 103-109.

(51) B. N. Ames, M. Profet, and L. S. Gold, *Proc. Natl. Acad. Sci. USA*, **1990**,*87*, 7777-7781.

(52) B. N. Ames, M. Profet, and L. S. Gold, *Proc. Natl. Acad. Sci. USA*, **1990**,*87*, 7782-7786.

(53) L. S. Gold, T. H. Slone, and B. N. Ames. In *Food Chemical Risk Analysis*; Tennant, D., ed.; Chapman & Hall Ltd: London, 1997; pp 267-295.

(54) B. N. Ames, L. S. Gold, and M. K. Shigenaga, *Risk Anal.*, **1996**,*16*, 613-617.

(55) L. S. Gold, T. H. Slone, and B. N. Ames. In *Handbook of Carcinogenic Potency and Genotoxicity Databases*; Gold, L.S., Zeiger, E., eds.; CRC Press: Boca Raton, FL, 1997; pp 661-685.

(56) B. N. Ames, and L. S. Gold. In *Risks, Costs, and Lives Saved: Getting Better Results from Regulation*; Hahn, R.W., ed.; Oxford University Press and AEI Press: New York and Washington, DC, 1996; pp 4-45.

(57) L. S. Gold, L. Bernstein, R. Magaw, and T. H. Slone, *Environ. Health Perspect.*, **1989**,*81*, 211-219.

(58) B. N. Ames, M. K. Shigenaga, and L. S. Gold, *Environ Health Perspect*, **1993**,*101(Suppl 5)*, 35-44.

(59) E. G. Luebeck, K. B. Grasl, T. I. Timmermann, W. Bursch, H. R. Schulte, and S. H. Moolgavkar, *Toxicol. Appl. Pharmacol.*, **1995**,*130*, 304-315.

(60) K. W. Kinzler, and B. Vogelstein, *Nature*, **1996**,*379*, 19-20.

(61) T. M. Monticello, J. A. Swenberg, E. A. Gross, J. R. Leininger, J. S. Kimbell, S. Seilkop, T. B. Starr, J. E. Gibson, and K. T. Morgan, *Cancer Res.*, **1996**,*56*, 1012-1022.

(62) G. Vomiero-Highton, and J. Heddle, *Mutagenesis*, **1995**,*10*, 381-384.

(63) B. N. Ames, and L. S. Gold, *Proc. Natl. Acad. Sci. USA*, **1990**,*87*, 7772-7776.

(64) J. R. M. Innes, B. M. Ulland, M. G. Valerio, L. Petrucelli, L. Fishbein, E. R. Hart, A. J. Pallota, R. R. Bates, H. L. Falk, J. J. Gart, M. Klein, I. Mitchell, and J. Peters, *J. Natl. Cancer Inst.*, **1969**,*42*, 1101-1114.

(65) T. S. Davies, and A. Monro, *J. Am. Coll. Toxicol.*, **1995**,*14*, 90-107.

(66) J. Contrera, A. Jacobs, and J. DeGeorge, *Regul. Toxicol. Pharmacol.*, **1997**,*25*, 130-145.

(67) D. L. Laskin, and K. J. Pendino, *Annu Rev Pharmacol Toxicol*, **1995**,*35*, 655-677.

(68) H. Wei, and K. Frenkel, *Carcinogenesis*, **1993**,*14*, 1195-1201.

(69) L. Wei, H. Wei, and K. Frenkel, *Carcinogenesis*, **1993**,*14*, 841-847.

(70) D. L. Laskin, F. M. Robertson, A. M. Pilaro, and J. D. Laskin, *Hepatology*, **1988**,*8*, 1051-1055.

(71) M. J. Czaja, J. Xu, Y. Ju, E. Alt, and P. Schmiedeberg, *Hepatology*, **1994**,*19*, 1282-1289.

(72) Y. Adachi, L. E. Moore, B. U. Bradford, W. Gao, and R. G. Thurman, *Gastroenterology*, **1995**,*108*, 218-224.

(73) L. Gunawardhana, S. A. Mobley, and I. G. Sipes, *Toxicol Appl Pharmacol*, **1993**,*119*, 205-213.

(74) B. Butterworth, R. Conolly, and K. Morgan, *Cancer Lett.*, **1995**,*93*, 129-146.

(75) M. L. Cunningham, J. Foley, R. Maronpot, and H. B. Matthews, *Toxicol. Appl. Pharmacol.*, **1991**,*107*, 562-567.

(76) M. L. Cunningham, and H. B. Matthews, *Toxicol. Appl. Pharmacol.*, **1991**,*110*, 505-513.

(77) M. L. Cunningham, M. R. Elwell, and H. B. Matthews, *Environ. Health Perspect.*, **1993**,*101(Suppl 5)*, 253-258.

(78) M. L. Cunningham, M. R. Elwell, and H. B. Matthews, *Fundamental and Applied Toxicology*, **1994**,*23*, 363-369.

(79) M. L. Cunningham, R. R. Maronpot, M. Thompson, and J. R. Bucher, *Toxicol. Appl. Pharmacol.*, **1994**,*124*, 31-38.

(80) J. Yarbrough, M. Cunningham, H. Yamanaka, R. Thurman, and M. Badr, *Hepatology*, **1991,***13*, 1229-1234.

(81) M. L. Cunningham, L. L. Pippin, N. L. Anderson, and M. L. Wenk, *Toxic. and Appl. Pharmacol.*, **1995,***131*, 216-223.

(82) J. Thottassery, L. Winberg, J. Youseff, M. Cunningham, and M. Badr, *Hepatology*, **1992,***15*, 316-322.

(83) J. Hayward, B. Shane, K. Tindall, and M. Cunningham, *Carcinogenesis*, **1995,***16*, 2429-2433.

(84) M. L. Cunningham, *Mutation Res.*, **1996,***365*, 59-69.

(85) R. J. Griffin, C. N. Dudley, and M. L. Cunningham, *Fundam. Appl. Toxicol.*, **1996,***29*, 147-154.

(86) J. C. Mirsalis, Provost, G.S., Matthews, C.D., Hamner, R.T., Schindler, J.E., O'Loughlin, K.G., MacGregor, J.T. and Short, J.M., *Mutagenesis*, **1993,***8*, 265-271.

(87) P. Shaver-Walker, C. Urlando, K. Tao, X. Zhang, and J. Heddle, *Proc. Natl. Acad. Sci. USA*, **1995,***92*, 11470-11474.

(88) J. Larson, D. Wolf, and B. Butterworth, *Fundam. Appl. Toxicol.*, **1994,***22*, 90-102.

(89) M. Ochiai, M. Watanabe, H. Kushida, K. Wakabayashi, T. Sugimura, and M. Nagao, *Carcinogenesis*, **1996,***17*, 95-98.

(90) A. Okumura, T. Tanaka, and H. Mori, *Jpn. J. Cancer Res.*, **1996,***87*, 805-815.

(91) S. Cohen, and T. Lawson, *Cancer Letters*, **1995,***93*, 9-16.

(92) S. M. Cohen, and L. B. Ellwein, *Cancer Res.*, **1991,***51*, 6493-6505.

(93) S. Cohen, *Regul. Toxicol. Pharmacol.*, **1995,***21*, 75-80.

(94) J. Counts, and J. Goodman, *Reg. Toxic. Pharm.*, **1995,***21*, 418-421.

(95) S. Preston-Martin, M. C. Pike, R. K. Ross, P. A. Jones, and B. E. Henderson, *Cancer Res.*, **1990,***50*, 7415-7421.

(96) S. Preston-Martin, K. Monroe, P.-J. Lee, L. Bernstein, J. Kelsey, S. Henderson, D. Forrester, and B. Henderson, *Cancer Epidemiol. Biomarkers*, **1995,***4*, 333-339.

(97) L. S. Gold, T. H. Slone, B. R. Stern, N. B. Manley, and B. N. Ames. In *Comparative Environmental Risk Assessment*; Cothern, C.R., ed.; Lewis Publishers: Boca Raton, FL, 1993; pp 209-235.

(98) B. N. Ames, R. Magaw, and L. S. Gold, *Science*, **1987,***236*, 271-280.

(99) L. S. Gold, T. H. Slone, N. B. Manley, and B. N. Ames, *Canc. Lett.*, **1994,***83*, 21-29.

(100) L. S. Gold, G. B. Garfinkel, and T. H. Slone. In *Chemical Risk Assessment and Occupational Health, Current Applications, Limitations, and Future Prospects*; Smith, C.M., Christiani, D.C., Kelsey, K.T., eds.; Greenwood Publishing Group: Westport, CT, 1994; pp 91-103.

(101) National Research Council, *Carcinogens and Anticarcinogens in the Human Diet: A Comparison of Naturally Occurring and Synthetic Substances*, National Academy Press, Washington, D.C., 1996.

(102) U. S. A. National Academy of Sciences, *The Life Sciences: Recent Progress and Application to Human Affairs, the World of Biological Research, Requirement for the Future*, Committee on Research in the Life Sciences, Washington, D.C., 526 p., 1970.

(103) J. H. Renwick, W. D. B. Claringbold, M. E. Earthy, J. D. Few, and A. C. S. McLean, *Teratology*, **1984,***30*, 371-381.

(104) T. Colburn, D. Dumanoski, and J. P. Myers. *Our Stolen Future: Are we Threatening our Fertility, Intelligence, and Survival?: A Scientific Detective Story.*, Dutton: New York, 1996.

(105) S. H. Safe, *Environ. Sci. Pollution Res.*, **1994,***1*, 29-33.

(106) S. H. Safe, *Env. Health Persp.*, **1995,***103*, 346-351.

(107) K. Reinli, and G. Block, *Nutr. Cancer*, **1996,***26*, 1996.

(108) G. Kolata, Measuring men up, sperm by sperm, The New York Times, May 4, E4(N), E4(L), (col.1), (1996).

(109) W. K. Viscusi. *Fatal Trade-offs*, Oxford University Press: Oxford, England, 1992.

(110) J. D. Shanahan, and A. D. Thierer. (1996) How to talk about risk: How well-intentioned regulations can kill: TP13, Washington, D.C., Heritage Foundation.

(111) *Risks, Costs, and Lives Saved: Getting Better Results from Regulation;* R. W. Hahn, ed.; Oxford University Press and AEI Press: New York and Washington, DC, 1996.

(112) *Risk versus Risk: Tradeoffs in Protecting Health and the Environment;* J. Graham, and J. Wiener, eds.; Harvard University Press: Cambridge, Massachusetts, 1995.

(113) J. Hadley, and A. Osei, *Medical Care*, **1982,***20*, 901-914.

(114) T. O. Tengs, M. E. Adams, J. S. Pliskin, D. G. Safran, J. E. Siegel, M. C. Weinstein, and J. D. Graham, *Risk Anal.*, **1995,***15*, 369-389.

(115) T. O. Tengs, and J. D. Graham. In *Risks, Costs, and Lives Saved: Getting Better Results from Regulation*; Hahn, R., ed.; Oxford University Press: New York, NY, 1996; pp 165-173.

(116) B. N. Ames, and L. S. Gold, *Science*, **1990,***249*, 970-971.

(116) B. N. Ames, and L. S. Gold, *FASEB J.*, **1997,** in press.

Chapter 2

Glucosinolates, Myrosinase, and Isothiocyanates: Three Reasons for Eating Brassica Vegetables

Jed W. Fahey, Katherine K. Stephenson, and Paul Talalay

[1]Brassica Chemoprotection Laboratory and Department of Pharmacology and Molecular Sciences, School of Medicine, Johns Hopkins University, 725 North Wolfe Street, Baltimore, MD 21205–2185

Numerous epidemiological studies provide consistent evidence that individuals who consume the highest quantiles of fruit and vegetables experience lower risks of developing several types of cancer (reviewed by *1-3*). Although cruciferous vegetables (and especially *Brassica* sp.) are believed to play a special role in this risk reduction (*4,5*), it is by no means clear how much of the protective effect can be attributed to the nutrient or the non-nutrient components of these foods, or to what extent indirect effects such as a parallel reduction in fat consumption and increases in fiber or vitamin and carotenoid intake may be responsible for protection. Epidemiological studies do, however, support the contention that dietary modification could reduce cancer risk by one third to one half in economically developed countries (*6,7*). Furthermore, it is believed that the majority of cancer cases (1.4 million new cases are expected in the U.S. alone in 1997) could be avoided by reducing various risk factors (tobacco, radiation, alcohol, exposure to certain drugs and chemicals) in combination with dietary modification (*8*). More recently, the prospect of reducing susceptibility to carcinogens by the administration of various chemical (including dietary) protectors has become widely recognized as both a feasible and a rational strategy in the battle against cancer (*9-12*). This approach has been termed chemoprotection, chemoprevention, or chemoprophylaxis.

One major strategy for achieving chemoprotection against cancer depends on the induction of Phase 2 detoxication enzymes in mammalian tissue by the administration of dietary or synthetic chemical agents (*13-15*). Although chronic exposure to various metabolic stresses cannot be always regarded as beneficial, exposure of animals and their cultured cells to low levels of certain of these stresses, especially electrophilic chemicals, causes an acceleration of reduced glutathione synthesis. In addition, electrophile exposure causes a coordinated induction of many of the Phase 2 enzymes that detoxify these compounds, thus reducing the susceptibility of cells to higher concentrations of the inducer as well as other related electrophiles (*16,17*). These Phase 2 enzymes (e.g. glutathione transferases, NAD(P)H:quinone oxidoreductase, glucuronosyltransferase, and epoxide hydrolase) inactivate ultimate carcinogens by destruction of their reactive centers or by conjugation with endogenous ligands thereby neutralizing their toxic properties and accelerating their elimination from the body. Evidence for a causal relationship between Phase 2 enzyme induction and protection has been mounting and can be considered as firmly established (14,15). Recently, considerable attention has been focused on edible plants which are rich in phytochemicals that induce Phase 2 enzymes (18-20). This approach is consistent with the epidemiological evidence which also strongly suggests the involvement of plants in cancer chemoprotection . Thus, cruciferous vegetables (e.g. broccoli, cauliflower, kale, turnips, collards, Brussels sprouts, cabbage, kohlrabi, rutabaga, Chinese cabbage, bok choy) contain water soluble secondary metabolites called glucosinolates, which are converted by endogenous enzymes (myrosinases) into highly reactive isothiocyanates as a defense response to predation or injury. The resultant isothiocyanates (R-N=C=S) or mustard oils are the principal Phase 2 enzyme inducers of cruciferous plants. The most potent naturally-occurring inducer isothiocyanate is sulforaphane (19). Sulforaphane was isolated in this laboratory from broccoli by monitoring the inducer activity of extracts of this plant. The inducer and anticarcinogenic properties of sulforaphane have also been demonstrated in animal tissues (21,22). The latent inducer activity of extracts of 3-day-old broccoli seedlings (sprouts) is attributable almost entirely to the presence of glucoraphanin (the glucosinolate precursor of sulforaphane). The concentration of inducer activity in such sprouts (200,000 - 800,000 units per gram fresh weight) is much higher than that of mature broccoli heads (5,000 - 80,000 units per gram fresh weight). When comparable doses of sulforaphane and of broccoli sprout extracts (based on inducer activity) were used to treat rats that also received single doses of the potent carcinogen DMBA (dimethylbenz[a]anthracene), a quantitatively similar reduction in mammary tumor development was observed (20).

The very stable glucosinolate precursors of isothiocyanates are typically present in some plants (e.g. crucifers) at much higher concentrations than their cognate isothiocyanates. Glucosinolate levels can amount to as much as 1% (w/w) in tissues of some cruciferous plant species. Glucosinolates comprise about 0.05 - 0.1% of the fresh weight of broccoli or about 50 - 100 mg of these compounds per 100 g portion. Glucosinolates are hydrolyzed to isothiocyanates by the coexisting enzyme myrosinase which is physically segregated within plant cells. It is commonly accepted that myrosinase activity is initiated when plants are damaged by insects or other predators, by food preparation, by chewing, or by other forms of damage, such as bruising or freeze-thawing during cultivation, harvest, shipping or handling, and in animal digestive systems. These types of tissue damage release myrosinase [thioglucohydrolase; EC 3.2.3.1] which cleaves the thioglucose linkage thereby releasing glucose. The resulting unstable intermediate undergoes [a probably nonenzymatic] rearrangement in which sulfate is released and isothiocyanates as well as

$$R-C \overset{\displaystyle S - \beta\text{-}D\text{-}C_6H_{11}O_5}{\underset{\displaystyle N\text{-}O\text{-}SO_3}{\Big\langle}} \xrightarrow{\text{MYROSINASE}} R-N=C=S \quad + \quad glucose \quad + \quad HSO_4^-$$

GLUCOSINOLATE   ISOTHIOCYANATE

other products are formed. Nearly all of the inducer activity in extracts of crucifers arises from glucosinolates which are not biologically active as such, but must undergo hydrolysis to isothiocyanates in order to manifest Phase 2 enzyme inducer activity (20).

Glucosinolates were first isolated in the middle of the last century and much effort has been devoted to developing methods for their efficient isolation and identification. Existing methods for separation involved ion exchange, GC, and HPLC, mostly after chemical modification (enzymatic sulfate removal and/or silylation of sugar moieties), but chromatographic standards are difficult to obtain and in many methods biological activity of the molecule is destroyed. We have therefore developed methods for the separation and identification of individual glucosinolates in plant extracts by use of paired-ion chromatography in the presence of tetraoctyl- or tetradecyl-ammonium bromide in combination with mass and NMR spectroscopy for final confirmation of identity (23). The paired-ion chromatography methods are extensions of methods developed by others (24,25).

Paired-ion chromatography of glucosinolates in conjunction with myrosinase hydrolysis and the quantitation of the resultant isothiocyanates by cyclocondensation with 1,2-benzenedithiol provides a powerful and comprehensive procedure for rapidly characterizing and quantitating glucosinolates in plant extracts (23). The cyclocondensation reaction with vicinal dithiols was developed for the quantitation of isothiocyanates. It exploits the ability of isothiocyanates to form cyclic thiocarbonyl derivatives with vicinal dithiols. When 1,2-benzenedithiol is the analytical reagent, the resultant 1,3-benzodithiole-2-thione has highly favorable properties for spectroscopic quantitation ($a_m = 23,000$ $M^{-1}cm^{-1}$ at $\lambda_{max} = 365$ nm)(26). The cyclocondensation procedure has been recently improved and can now be used to measure as little as 10 picomoles of isothiocyanates in complex biological mixtures such as plant extracts (27). In all cases examined, inducer activity of broccoli extracts correlated with their potential isothiocyanate content (i.e., endogenous isothiocyanate plus isothiocyanate released by the action of an excess of added purified myrosinase).

Phase 2 enzyme inducer potency varies according to a number of factors. We have obtained preliminary data showing that there are large cultivar effects as well as tremendous environmentally induced variation due to cultivation, handling and storage parameters. Once tissue is damaged (e.g. at harvest, or due to improper handling), the enzyme myrosinase initiates the conversion of relatively non-reactive glucosinolates to isothiocyanates and other breakdown products which are more highly reactive than glucosinolates. There is thus an immediate opportunity for loss of Phase 2 inducer potency since glucosinolate breakdown products such as isothiocyanates initiate further degradative reactions within the plant tissue. A sampling of fresh broccoli collected randomly from 22 Baltimore area supermarkets had qualitatively similar glucosinolate profiles, but had a greater than 8-fold range of inducer potencies between the highest and lowest sample (20). This tremendous variability in glucosinolate content [and hence Phase 2 inducer potency] thus translates to an inability of the consumer to rationally select broccoli for inducer potency using criteria such as appearance and odor for there are no such clues as to their ranking.

The common alkylthioalkyl glucosinolates (e.g. glucoraphanin, glucoiberin and glucoerucin; see Figure 1) form isothiocyanates (sulforaphane, iberin and erucin

Glucoraphanin
(4-methylsulfinylbutyl)

$$CH_3-\overset{\displaystyle O}{\underset{\displaystyle \|}{S}}-(CH_2)_4-R*$$

Glucoiberin
(3-methylsulfinylpropyl glucosinolate)

$$CH_3-\overset{\displaystyle O}{\underset{\displaystyle \|}{S}}-(CH_2)_3-R$$

Glucoerucin
(4-methylthiobutyl glucosinolate)

$$CH_3-S-(CH_2)_4-R$$

Glucobrassicin
(indole-3-ylmethyl glucosinolate)

Neoglucobrassicin
(1-methoxyindole-3-ylmethyl glucosinolate)

4-Hydroxyglucobrassicin
(4-hydroxyindole-3-ylmethyl glucosinolate)

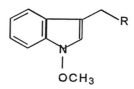

$$*R= \quad -C\overset{\displaystyle S-C_6H_{11}O_5}{\underset{\displaystyle N-OSO_2O^-}{}}$$

Figure 1. Glucosinolates commonly found in broccoli.

respectively) with very low cytotoxicity and extremely high Phase 2 enzyme inducing potency. In fact sulforaphane is the most potent naturally occurring Phase 2 enzyme inducer. These compounds are monofunctional inducers since they do not also induce Phase 1 enzymes which can activate xenobiotics thus potentially generating carcinogens (*19,28*).

Indole glucosinolates (e.g. glucobrassicin, neoglucobrassicin, 4-hydroxyglucobrassicin, 1-hydroxyglucobrassicin and 4-methoxyglucobrassicin) do not form stable isothiocyanates when hydrolysed by myrosinase. Unstable intermediates are formed in the enzymatic reaction and give rise to compounds such as indole-3-carbinol, indole-3-acetonitrile and 3,3'-diindolylmethane (*29-31*). These compounds are only weak inducers of Phase 2 detoxification enzymes (unpublished obervations). For example, the inducer potency of indole-3-acetonitrile is over 200 times lower than that of sulforaphane. However, these compounds also induce Phase 1 enzymes, i.e. they are bifunctional inducers (*28*), so that they can participate in the formation of carcinogens by modulating the metabolic activation of procarcinogens. Furthermore, in the acid conditions in the stomach, indole-3-carbinol can also spontaneously condense to cyclic structures which resemble TCDD (dioxin), and bind with high affinity to the Aryl hydrocarbon (Ah) receptor, and induce certain cytochromes P-450 that can activate procarcinogens. Metabolic derivatives of indole glucosinolates may therefore act simultaneously both to promote tumors and to prevent their initiation (*30,32-34*).

Myrosinase hydrolysis of another class of glucosinolates, the β-hydroxyalkenyl glucosinolates (e.g. progoitrin and epiprogoitrin), gives rise to intermediates that apparently cyclize spontaneously to goitrogenic oxazolidonethiones. Over-consumption of certain *Brassica* vegetables such as kale and cabbage has been known for many years to cause goiter in both experimental animals and humans and may be responsible for the development of endemic goiter in certain regions (*35*). Although some crucifers such as cabbage and especially the oilseed crops crambe and rapeseed contain large quantities of these goitrogenic glucosinolates, broccoli does not contain significant quantities of these compounds.

Efforts are being made to understand the complex effects of diet on cancer incidence. We have focused on minor dietary constituents that elevate the activity of cellular detoxication or Phase 2 enzymes based on evidence indicating that induction of these enzymes blocks the formation of tumors in experimental animals. Sensitive analytical techniques have been developed for the quantitation of glucosinolates and of isothiocyanates and these methods are being used to evaluate the anticancer potential of a range of cruciferous vegetables and the environmental and genetic components of glucosinolate production. Since glucosinolates are water-soluble, cooking and other forms of processing should be performed, if at all, in such a way (e.g. by steaming, microwaving, stir-frying or rapid boiling with minimal water) as to avoid excessive leaching of the compounds.

The utility of these plants in selectively elevating Phase 2 enzymes (and not Phase 1 enzymes) is a promising facet in the rational development of dietary strategies for reducing the risk of cancer.

Acknowledgments

These studies were supported by gifts from Lewis B. Cullman, Charles B. Benenson, and other Friends of the Brassica Chemoprotection Laboratory, by grants from the American Institute for Cancer Research, the Cancer Research Foundation of America and a program project grant (1PO1 CA44530) from the National Cancer Institute, Department of Health and Human Services.

Literature Cited

1. Block, G.; Patterson, B.; Subar, A. *Nutr Cancer* **1992**, *18*; 1-29.
2. Steinmetz, K.; Potter, J. *J. Amer. Diet. Assoc.* **1996**, *96*; 1027-1039.
3. *Food, Nutrition and the Prevention of Cancer: A Global Perspective;* Potter, J.D. Ed.; World Cancer Research Foundation - American Institute for Cancer Research; Washington, D.C., 1997; p630.
4. Hertog, M.G.L; Bueno-de-Mesquita, H.B.; Fehily, A.M.; Sweetnam, P.M.; Elwood, P.C.; Kromhout, D. *Cancer Epidemiology, Biomarkers & Prevention* *1996*, 5; 673-677.
5. Verhoeven, D.T.H.; Goldbohm, R.A.; vanPoppel, G.; Verhagen, H.; van den Brandt, P.A. *Cancer Epidemiology, Biomarkers & Prevention* **1996**, *5*; 733-748.
6. Willett, W.C.; Trichopoulos, D. *Cancer Causes and Control* **1996,** *7*; 178-180.
7. Doll, R. *Cancer Res.* **1992**, *52*; 2024s-2029s.
8. Doll, R.; Peto, R. *J. Natl. Cancer Inst.* **1981**, *66*; 1191-1308.
9. Wattenberg, L.W. *Cancer Res.* **1992,** *52*; 2085s-2091s.
10. Wattenberg, L.W. *Preventive Medicine* **1993**, 25; 44-45.
11. Wattenberg, L.W. *Ann. N.Y. Acad. Sci.* **1995,** *768*; 73-81.
12. Kelloff, G.J.; Boone, C.W.; Crowell, J.A.; Steele, V.A.; Lubet, R.A.; Doody, L.A.; Malone, W.F.; Hawk, E.T.; Sigman, C.C. *J. Cell. Biochem. - Supp.* **1996,** *26*; 1-28.
13. Prochaska H.J.; Talalay, P. In *Phenolic Compounds in Food and Their Effects on Health II;* Huang, M.-T.; Ho, C.-T.; Lee, C.Y., Eds.; Amer. Chem. Soc. Sympos. Ser. 507; Washington D.C., 1992; pp 150-159.
14. Talalay, P.; Fahey, J.W.; Holtzclaw, W.D.; Prestera, T.; Zhang, Y. *Toxicology Letters* **1995,** *82/83*; 173-179.
15. Talalay, P.; Zhang, Y. *Biochem. Soc. Trans.* **1996,** *24*; 806-810.
16. Kahl, R. *Toxicology* **1984**, 33; 185-228.
17. Prestera, T.; Zhang, Y.; Spencer, S.R.; Wilczak, C.A.; Talalay, P. *Advan. Enzyme Regul.* **1993**, 33; 281-296.
18. Prochaska, H.J.; Santamaria, A.B.; Talalay, P. *Proc. Natl. Acad. Sci. USA* **1992,** *89*; 2394-2398.
19. Zhang, Y.; Talalay, P.; Cho, C.-G.; Posner, G.H. *Proc. Natl. Acad. Sci. USA* **1992,** *89*; 2399-2403.
20. Fahey, J.W.; Zhang, Y.; Talalay, P. *Proc. Natl. Acad. Sci. USA* **1997,** *94*; 10367-10372.
21. Zhang, Y.; Kensler, T.W.; Cho, C.-G.; Posner, G.H.; Talalay, P. *Proc. Natl. Acad. Sci. USA* **1994,** *91*; 2147-2150.

22. Jang, M.; Cai, L.; Udeani, G.O.; Slowing, K.W.; Thomas, C.F.; Beecher, C.W.; Fong, H.H.; Farnsworth, N.R.; Kinghorn, A.D.; Mehta, R.G.; Moon, R.C.; Pezzuto, J.M. *Science* **1997**, *275(5297)*; 218-20.

23. Prestera T.P.; Fahey, J.W.; Holtzclaw, W.D.; Abeygunawardana, C.; Kachinski, J.L.; Talalay, P. *Anal. Biochem.* **1996**, *239*; 168-179.

24. Betz, J.M.; Fox, W.D. In *Food Phytochemicals for Cancer Prevention. I. Fruits and Vegetables*; Huang, M.T.; Osawa, T.; Ho, C.-T.; Rosen, R.T., Eds.; ACS Symposium Series 546; Am. Chem. Soc., Washington, D.C., 1994, Vol 546; pp. 181-196.

25. Helboe, P.; Olsen, O.; Sørensen, H. *J. Chromatogr.* **1980**, *197*; 199-205.

26. Zhang, Y.; Cho, C.-G.; Posner, G.H.; Talalay, P. *Anal. Biochem.* **1992**, *205*; 100-107.

27. Zhang, Y.; Wade, K.L.; Prestera, T.; Talalay, P. *Anal. Biochem.* **1996**, *239*; 160-167.

28. Prochaska, H.J.; Talalay, P. *Cancer Research* **1988**, *48*; 4776-4782.

29. Bradfield, C.A.; Bjeldanes, L.F. *J. Toxicol. Environ. Health* **1987**, *21*; 311-323.

30. Kim, D.J.; Han, B.S.; Ahn, B.; Hasegawa, R.; Shirai, T.; Ito, N.; Tsuda, H. *Carcinogenesis* **1997**, *18(2)*; 377-381.

31. Ames, B.N.; Profet, M.; Gold, L.S. *Proc. Nat. Acad. Sci. USA* **1990**, *87*; 7782-7786.

32. Bjeldanes, L.F.; Kim, J.-Y.; Grose, K.R.; Bartholomew, J.C.; Bradfield, C.A. *Proc. Natl. Acad. Sci. USA* **1991**, *88*; 9543-9547.

33. Bailey, G.S.; Hendricks, J.D.; Shelton, D.W.; Nixon, J.E.; Pawlowski, N.E. *JNCI* **1987**, *78(5)*; 931-934.

34. Pence, B.C.; Buddingh, F.; Yang, S.P. *JNCI* **1986**, *77(1)*; 269-276.

35. Michajlovskij, N.; Sedlak, J.; Jusic, M.; Buzina, R. *Endocrinol Exper.* **1969**, *3*; 65-72.

Chapter 3

Chemoprevention of Colon Cancer by Select Phytochemicals

B. S. Reddy

Division of Nutritional Carcinogenesis, American Health Foundation, Valhalla, NY 10595

Cancer of the colon is the leading cause of cancer deaths in the United States and western countries. Epidemiological data, including both cohort and case-control studies suggest that plant foods including fruits and vegetables have preventive potential. Plant foods are a good source of potentially antimutagenic and anticarcinogenic agents including carotenoids, allium compounds, and curcuminoids, to cite a few. Laboratory studies have provided relevant mechanistic and efficacy data on the role of specific phytochemicals in the carcinogenic process. Dietary administration of diallyl disulfide, anetholetrithione, curcumin, caffeic acid esters, and phytic acid inhibits colon carcinogenesis in the rodent model. Mechanistically, organosulfur compounds have been found to modulate the formation and bioactivation of carcinogens whereas curcumin and caffeic acid esters have been shown to induce phase II enzymes and inhibit cyclooxygenase and/or lypoxygenase activities suggesting that the modulation of arachidonic acid by these agents may play a role in their chemopreventive activity.

Several human epidemiological studies and supporting data from laboratory animal studies suggest that dietary factors play an important role in the etiology of several cancers. Epidemiological studies, both cohort and case-control, strongly suggest that consumption of certain vegetables and fruits reduce the risk of development of cancers at many sites including colon (*1*). Among specific categories of vegetables, higher intake of cruceferous vegetables was associated with lower risk for colon cancer development (*1*). There are many biologically plausible reasons as to why dietary intake of plant foods might slow or prevent the occurrence of cancer. Fruits and vegetables are a good source of potentially anticarcinogenic and antimutagenic substances including carotenoids, vitamin E, selenium, dietary fiber, dithiolethiones, protease inhibitors, oligofructoses, allium compounds, and curcuminoids, to cite a few. Inhibition of development of cancer by administration of dietary factors that directly and/or indirectly inhibit the cancer-producing effects of genotoxic and promoting substances offers an excellent approach to cancer prevention. As our understanding of the mechanisms of carcinogenesis has increased, possibilities have emerged for interfering at multiple points along carcinogenic process namely during initiation, promotion and progression. Chemoprevention refers to the administration of chemical agents, both naturally occurring micronutrients and non-nutrients in foods and synthetic, to prevent the initiational and promotional events occurring during the process of neoplastic development (2,3). Wattenberg (3) has classified chemopreventive agents into three broad categories, with distinctly different functions; these include the agents that can block the metabolic activation of carcinogens; agents that can prevent the formation of carcinogens from precursors; and agents that can suppress the expression of neoplasia in cells previously exposed to an effective dose of carcinogen. In addition, several chemopreventive agents have been shown to inhibit tumorigenesis both by blocking the metabolic activation of carcinogens and by suppressing the promotion and progression of tumors. Several micronutrients and non-nutrients present in food fit very well into these categories. The great diversity of the compounds is a positive feature indicating the possibility that a variety of approaches to cancer prevention by these agents can be made and that the options for selecting the right chemopreventive agents will be large (3).

Naturally-occurring non-nutrient substances that have been intensely investigated for their potential chemopreventive properties include organosulfur compounds, aromatic isothiocyanates, dithiolethiones, polyphenols, curcumin, phytic acid, and ellagic acid, to cite a few. This chapter does not attempt to review all compounds present in food but focuses on few a agents that have been shown to inhibit colon carcinogenesis.

Organosulfur Compounds

Diallyl Disulfide. Organosulfur compounds present in garlic and onions, such as allyl sulfide, allyl disulfide, and allyl methyl di- and trisulfides, and garlic and onion oils have been found to inhibit carcinogenesis at several organ sites (4,5). Allyl methyl trisulfide, diallyl sulfide, and diallyl trisulfide administered orally prior to carcinogen treatment (initiation stage) inhibited benzo[a]pyrene[B(a)P]-induced and nitrosodiethylamine-induced forestomach tumors in mice (4). In these studies, the compound to most inhibit tumorigenesis was diallyl disulfide. Also, diallyl disulfide administered during the initiation phase inhibited N-nitrosomethylbenzylamine-induced esophageal tumorigenesis in rats (5).

The effect of diallyl disulfide on colon carcinogenesis was also investigated in the laboratory animal models (6,7). Diallyl disulfide administered in the diet prior to carcinogen treatment inhibited 1,2-dimethylhydrazine (DMH)-induced colon carcinogenesis in rats. The inhibitory effects of these compounds during the initiation phase may be due to inhibition of carcinogen activation through the suppression of oxidative metabolism of carcinogen. Administration of diallyl disulfide during the postinitiation phase inhibited DMH-induced colon carcinogenesis in rats, but this inhibition of colon carcinogenesis was associated with the retardation of body weight gain.

We have investigated the effect of dietary diallyl disulfide on azoxymethane (AOM)-induced colon carcinogenesis in male F344 rats (7). Also, the effects of this agent on the activities of phase II enzymes, namely glutathione s-transferase (GST), NAD(P)H-dependent quinone reductase (QR), and UDP-glucuronsosyl transferase (UDP-GT) in the colon were determined. Prior to initiation of chemoprevention study, the maximum tolerated dose (MTD) level of diallyl disulfide was determined in a subchronic toxicity study. The MTD of diallyl disulfide was found to be 250 ppm. Dietary administration of 100 and 200 ppm diallyl disulfide equivalent to 40 and 80% MTD levels significantly inhibited the invasive adenocarcinomas of the colon (Table I). The inhibition of colon carcinogenesis by diallyl disulfide is associated with an increase in the activities of detoxifying enzymes such as GST, QR and UDP-GT in the colonic mucosa and tumors (Table II). It is possible that the protective effect of diallyl disulfide may be accounted for, at least in part, by its ability to induce the detoxifying enzymes in the colon. However, the mechanism of chemopreventive activity of this agent administered during the postinitiation period is not clearly known.

Anethole Trithione. Several studies indicate that organosulfur compounds present in cruceferous vegetables (Brussels sprouts, cauliflower and cabbage) such as dithiolethiones have a role in cancer inhibition (3). Anethole trithione, a substituted dithiolthione has been evaluated in our laboratory for its chemopreventive activity against AOM-induced colon carcinogenesis (7). The MTD level of anethole trithione was found to be 250 ppm in male F344 rats. Administration of 200 ppm of anethole trithione in the diet significantly suppressed the incidence (percentage of animals with tumors) and multiplicity (tumors/animal) of both invasive and non-invasive adenocarcinomas of the colon whereas feeding of 100 ppm anethole trithione inhibited the invasive adenocarcinomas of the colon (Table I). Administration of anethole trithione strikingly increased the activities of GST, QR, and UDP-GT in the colonic mucosa and tumors (Table II). Although the exact mechanisms involved in its protective effects against colon carcinogenesis are not clearly elucidated, it appears that anethole trithione may, in part, inhibit colon carcinogenesis by blocking the formation of or trapping ultimate carcinogen electroophiles and/or by decreasing activation of carcinogen and by suppressing the effect of promoters in the target organ (3). The results of this study should provide a stimulus to design additional studies on the mechanisms of colon tumor inhibition by organosulfur compounds during the postinitiation stage of carcinogenesis.

Table I. Chemopreventive Properties of Organosulfur and Antiinflammatory Compounds Against Colon Carcinogenesis in Male F344 Rats[a]

Chemopreventive agent	Adenocarcinoma incidence (percent animals with tumors)			Adenocarcinomas/animal
	Invasive	Noninvasive	Total	
Experiment 1				
control diet	34	53	78	1.66±0.21[b]
100 ppm diallyl disulfide	11[c]	53	64	1.33±0.22
200 ppm diallyl disulfide	5.5[c]	58	64	1.10±0.21[c]
100 ppm anethole trithione	11[c]	47	58	1.00±0.18[c]
200 ppm anethole trithione	8.3[c]	30[c]	42[c]	0.64±0.15[c]
Experiment 2				
controld diet	38	67	81	1.50±1.04
2000 ppm Curcumin	20	37[c]	47[c]	0.66±0.54[c]
750 ppm PEMC	13[c]	40[c]	43[c]	0.70±0.98[c]

[a] Reddy et al. (7); Rao et al. (15). Five-week old male F344 rats were fed the control diet and experimental diets containing various chemoprevetnive agents. Two weeks later, all animals were given s.c. injections of azoxymethane (AOM) at a dose rate of 15 mg/kg body weight, once weekly for 2 weeks. All animals were continued on their respective dietary regimen until the termination of the experiment 52 weeks after the AOM treatment.

[b] Mean ± SD (n=36)

[c] Significantly different from the control diet group, $p < 0.05$

Table II. Effect of Dietary Organosulfur Compounds on Colonic Mucosal and Tumor on Phase II Enzyme Activity[a]

Chemopreventive agents	Colon mucosa			Colon tumors		
	GST[b]	NAD(P)H:QR	UDP-GT	GST	NAD(P)H:QR	UDP-GT
control diet	372±50[c]	450±57	1.7±0.36	209±44	336±38	1.8±0.2
200ppm diallyl disulfide	1089±174[d]	660±49[d]	2.0±0.2	960±117[d]	642±54[d]	2.5±0.4[d]
200 ppm anethole trithione	2010±262[d]	2702±381[d]	3.6±0.6[d]	1302±171[d]	2501±421[d]	3.7±0.6[d]

[a]Reddy et al. (7)

[b]GST activity is expressed as μmol 1-chloro-2,4-dinitrobenzene conjugated/mg protein/min; NAD(P)H:QR activity is expressed as μmol 2,6-dichloroindophenol reduced/mg protein/min; and UDP-GT activity is expressed as nmol p-nitrophenol conjugated/mg protein/min.

[c]Mean ± SD (n=6)

[d]Significantly different from the control diet group, p<0.01

Naturally Occurring Anti-inflammatory Agents in Foods

Curcumin. Turmeric, the powdered rhizome from the root of the plant *Curcumin longa Linn.*, has long been used as a spice for imparting color and flavor in foods and as a medicine for the treatment of inflammatory diseases. Curcumin (diferuloylmethane), a phenolic compound that has been identified as the major pigment in turmeric, possesses both anti-inflammatory (*9*) and antioxidant (*10*) properties. Previous studies have shown that topical application of curcumin inhibits 12-*O*-tetradecanoylphorbol-13-acetate (TPA)-induced epidermal DNA synthesis, tumor promotion in mouse skin, and edema of mouse ears (*11*). Topical application of curcumin has also inhibited B[*a*]P-induced DNA adducts and skin tumors as well as 7,12-dimethylbenz[*a*]anthracene (DMBA)-induced skin tumors in laboratory animal models (*12*). In other studies, dietary administration of 2% turmeric in the diet inhibited DMBA-induced skin, B[*a*]P-induced forestomach, and AOM-induced small and large intestinal tumors in mice (*13,14*). Studies from our laboratory showed that 0.2% curcumin in the diet significantly suppressed AOM-induced formation of colonic aberrant crypt foci, which are early preneoplastic lesions, in male F344 rats (*15*). In addition, administration of curcumin in a semipurified diet at 2000 ppm concentration to male F344 rats significantly inhibited the incidence of colon adenocarcinomas and the multiplicity of invasive and noninvasive adenocarcinomas induced by AOM (*15*; Table I). Dietary curcumin also significantly suppressed the colon tumor volume by >57%.

Experimental studies have shown that curcumin inhibits carcinogenesis by several mechanisms. With respect to mode of action during the initiation phase of carcinogenesis, curcumin has been shown to inhibit the metabolic activation of B[*a*]P-DNA adducts in the skin, forestomach and liver of mice (*16*). With regard to mode of action during the postinitiation stage, the results of our study demonstrated that curcumin suppressed the colonic mucosal and tumor phospholipase A_2 (PLA_2), phospatidylinositol-specific phospholipase C (PI-PLC), cyclooxygenase activities (formation of prostaglandins such as PGE_2, $PGF_2\alpha$, PGD_2 and 6-keto $PGF_1\alpha$, and thromboxane B_2, through the cyclooxygenase system) and lipoxygenase activity (production of 5(*S*)-, 8(*S*)-, 12(*S*)- and 15(*S*)-hydroxyeicosatetraenoic acids by lipoxygenase), suggesting that the modulation of arachidonic metabolism in the colon by curcumin may play a role in its inhibitory action against colon carcinogenesis (*15*; Table III). These metabolites of arachidonic acid exert a variety of biological activities. Cyclooxygenase metabolites, particularly prostaglandins of type-2 series modulate cell proliferation, tumor growth, and immune responses (*17,18*), whereas lypoxygenase metabolites can influence various biological responses including tumor cell adhesion and regulation of tumor metastatic potential (*19,20*). Additional studies conducted in our laboratory indicate that curcumin increases apoptosis in the colon tumors (*21*).

Table III. Effect of Curcumin and PEMC on Colonic Mucosal and Tumor Cyclo-oxygenase and Lipoxygenase Activity

	Control diet[a,b]		Curcumin[a] (2000 ppm in diet)		PEMC[b] (750 ppm in diet)	
	Mucosa	Tumor	Mucosa	Tumor	Mucosa	Tumor
Cyclo-oxygenase activity[c]						
PGE_2	328±13	1578±35	184±9[e]	789±35[e]	302±15	1178±32
$PGF_2\alpha$	355±16	937±29	236±14[e]	543±33[e]	328±13	694±22
PGD_2	242±13	549±20	156±6[e]	375±22[e]	248±16	423±21
6-keto $PGF1\alpha$	378±14	1193±46	284±13	820±29[e]	336±14	847±33[e]
thromboxane B_2	258±12	972±33	264±13	788±33[e]	254±14	825±27
Lipoxygenase activity[d]						
5(S)-HETE	211±12	312±13	93±6[e]	158±10[e]	148±9[e]	228±12
8(S)-HETE	312±21	348±16	176±12[e]	224±13[e]	165±12[e]	137±11[e]
12(S)-HETE	266±14	585±28	140±7[e]	253±21[e]	122±10[e]	213±14[e]
15(S)-HETE	308±22	428±31	189±16[e]	212±18[e]	204±13[e]	196±13[e]

[a]Rao et al., (15).
[b]Rao et al., (27)
[c]pmoles of prostaglandins and thromboxane B_2 formed (mean ± SEM) from[14C]arachidonic acid/mg protein/15 min.
[d]pmoles of HETEs (hydroxyeicosatetraenoic acids) formed (mean ± SEM) from [14C]arachidonic acid/mg protein/min.
[e]Significantly different from the control diet group; p<0.05

Caffeic Acid Esters. Propolis from honey beehives contains various chemical constituents that exhibit a broad spectrum of activities including antibacterial, antifungal, cytostatic and anti-inflammatory properties (22,23). Gribel and Pashinskii (24) have shown that honey possesses moderate antitumour and pronounced antimetastatic effects in tumors in five different strains of rats and mice. Caffeic acid (3,4-dihydroxycinnamic acid) and its esters, are present in propolis at levels of 20-25%, which are suspected of having a broad spectrum of biological activities including tumor suppression. Because of this potential, several caffeic acid esters that are present in propolis have been examined for their antimutagenic and antitumorigenic activities. The results of these studies showed that these agents inhibit colon adenocarcinoma HT-29 and HCT-116 cell growth (25). Among the caffeic acid esters evaluated, phenylethyl-3-methylcaffeate (PEMC) was found to be most effective in inhibiting aberrant crypt foci formation (26). In addition dietary administration of PEMC significantly inhibited the incidence and multiplicity of invasive, noninvasive and total (invasive plus noninvasive) adenocarcinomas of the colon (27; Table I) Dietary PEMC also suppressed the colon tumor volue by 43% compared with controls. This inhibition is mediated through the suppression of PI-PLC and lipoxygenase activities in colonic mucosa and tumor tissues. Although the exact mechanisms of chemopreventive action of PEMC remain to be elucidated, modulation of PI-PLC-dependent signal transduction pathways and eicosanoid metabolism as well as lipoxygenase activity (27; Table III) by PEMC probably play a role in its inhibitory action.

Inositol and Related Compounds

Inositol hexaphosphate (InsP$_6$, phytic acid) is a naturally occurring compound found in substantial amounts in cereals and legumes (28). In studies carried out to date, InsP$_6$ and inositol have shown chemopreventive activity not only against cancers of colon and mammary gland but also against cancers of liver, lung and skin (28). In separate studies, oral administration of InsP$_6$ was shown to inhibit colon carcinogenesis in rodents during the initiation and the postinitiation stages (28). Vucenik et al. (29) studied the chemopreventive properties of InsP$_6$ and inositol administered individually and in combination during initiation and postinitiation stages of DMBA-induced mammary carcinogenesis in Sprague Dawley rats. Animals supplemented with 15 mM InsP$_6$ or inositol, or 15 mM InsP$_6$ together with 15 mM inositol, in drinking water exhibited a significant reduction in tumor incidence and tumor burden. The combination of InsP$_6$ and inositol produced better chemopreventive activity than that of the compounds administered individually.

The exact mechanisms by which InsP$_6$ and inositol exert their chemopreventive effects have not been clearly demonstrated. InsP$_6$ has been shown to act as an antioxidant, to control cell division and reduce the rate of cell proliferation, and to enhance the activity of natural killer cells, which play an important role in the host defense against neoplasia (28).

Summary and Conclusions

Most human epidemiological data suggest that the majority of cancer deaths are attributable to lifestyle factors including nutritional factors. Diet, as it relates to specific food items, nutrients and non-nutrients, probably accounts for several major cancers including those of colon and breast. There is a reasonably consistent body of evidence from epidemiological, mechanistic, and preclinical efficacy studies that several minor non-nutrients present in fruits and vegetables may protect against several types of cancer. Given the continuing cancer burden and the relatively slow impact of proven cancer treatment strategies in reducing cancer mortality, it is important that there be an ongoing evaluation of promising minor non-nutrients present in foods as chemopreventive agents. Chemoprevention of cancer focuses on individuals who are at increased risk for cancer development. An element of importance in both identifying individuals at increased risk and conducting intervention trials is the use of intermediate biomarkers, which may help circumvent the length and expense involved in chemoprevention trials. The modulation of surrogate end points by minor non-nutrients in foods may serve as the rationale for initiating further clinical trials with cancer incidence as the end point.

Literature Cited

1.   Potter, J.D.; Steinmetz, K. *Principale of Chemoprevention*; Stewart, B.W.; McGregor, D.; Kleihues, P.; Eds.; IARC Scientific Publication No. 139; International Agency for Research on Carncer: Lyon, France 1966; 61-90.

2.   Kelloff, G.J.; Boone, C.W.; Crowell, J.A.;l Steele, V.E.; Lubet, R.; Sigman, C.C. Cancer Epidemiol Biomarkers Prev. 1994, *3*, 85-98.

3.   Wattenberg, L.W. In *Cancer Chemoprevention*; Wattenberg, L.W.; Lipkin, M.; Boone, C.W.; Kelloff, G.J.; Eds.; CRC PRess, Inc.: Boca RAton, FL, 1992; pp 19-39.

4.   Sparnins, V.L.; Barany, G.; Wattenberg, L.W. *Carcinogenesis* 1988, *9*, 131-134.

5.   Wargovich, M.J.; Woods, C.; Eng, V.W.S.; Stephens, L.C. *Cancer Res.* 1988, *48*, 6872-6875.

6.   Wargovich, M.J. *Carcinogenesis* 1987, *8*, 487-489.

7.   Reddy, B.S.; Rao, C.V.; Rivenson, A.; Kelloff, G. *Cancer Res.* 1993, 53: 3493-3498.

8.   Takahashi, S.; Hakoi, K.; Yada, H.; Hirose, M.; Ito, N.; Fukushima, S. *Carcinogenesis* 1992, *13*, 1513-1518.

9.   Srimal, R.X.; Dhawan, B.N. *J. Pharm. Pharmacol.* 1973, *25*, 447-452.

10. Toda, S.; Miyase, T.; Arichi, H.; Tanizawa, H.; Takino, Y. *Chem Pharm Bull.* 1985, *33*, 1726-1728.

11. Huang, M-T.; Smart, R.C.; Wong, G-Q.; Conney, A.H. *Cancer Res.* 1986, *48*, 5941-5946.

12. Huang, M-T.; Wang, Z.Y.; Georgiadis, C.A.; Laskin, J.D.; Conney, A.H. *Carcinogenesis.* *13*, 2183-2186.

13. Azuine, M.A.; Bhide, S.V. *Nutr. Cancer* 1992, *17*, 77-83.

14. Huang, M.-T.; Ma, W.; Low, Y-R.; Ferraro, T.; Reuhl, K.; Newmark, H.; Cooney, A.H. *Proc. Am. Assoc. Cancer Res.* *34*, 55.

15. Rao, C.V.; Desai, D.; Rivenson, A.; Simi, B.; Reddy, B.S. *Cancer Res.* 1995, *55*, 259-266.

16. Stoner, G.; Mukhtar, H. *J. Cell. Biochem.* 1995, *22*, 169-180.

17. Marnett, L.J.; *Cancer Res,* 1992, *52*, 5575-5589.

18. Smith, W.L.; *Am. J. Physiol.* 1992, *263*, F181-F191.

19. Honn, K.V.; Tang, D.G. *Cancer Metastasis Rev.* 1992, *11*, 353-375.

20. Timer, J.; Chen, Y.Q.; Linn, B.; Bazer, R.; Taylor, J.D.; Honn, K.V. *Int. J. Cancer.* 1992, *52*, 594-603.

21. Samaha, H.S.; Kelloff, G.J.;l Steele, V.; Rao C.V.; Reddy, B.S. *Cancer Res.* 1997, *57*, 1301-1305.

22. Jeddar A.; Khasany, A.; Ramsaroop, V.G.; Bhamjei, A.; Haffejce, I.E.; Moosa, A. *S. Afr. Med., J.* 1985, *67*, 257-258.

23. Koshihara, Y.; Neichi, T.; Murota, S.L.; Lao, A.; Fujimoto, Y.; Tatsuno, T. *Biochim, Biophys. Acta* 1984, *792*, 92-92.

24.   Gribel, N.V.; Pashinskii, V.G. *Drug Metab. Rev.* 1990, *36*, 704-709.

25.   Rao, C.V.; Desai, D.; Kaul, B.; Amin, S.; Reddy, B.S. *Chem. Biol. Interact.* 1992, *84*, 277-290.

26.   Rao, C.V.; Desai, D.; Simi, B.; Kulkarni, N; Amin, S.; Reddy, B. *Cancer Res.* 1993, *53*, 4182-4188.

27.   Rao, C.V.; Desai, D.; Rivenson, A.; Simi, B.; Amin, S.; Reddy, B.S. *Cancer Res.* 1995, *55*, 2310-2315.

28.   Shamsuddin, A.K. *J. Nutr.* 1995, *125*, 725S-732S.

Chapter 4

Lycopene and Cancer: An Overview

Joseph Levy, Michael Danilenko, Michael Karas, Hadar Amir, Amit Nahum, Yudit Giat, and Yoav Sharoni

Department of Clinical Biochemistry, Faculty of Health Sciences, Ben-Gurion University of the Negev and Soroka Medical Center of Kupat Holim, Beer-Sheva 84105, Israel

The remarkable association between diets rich in fruits and vegetables and the reduced risk of several malignancies has led to a consideration of the role of carotenoids in this context. Lycopene is one of the major carotenoids in Western diets and accounts for about 50% of carotenoids in human serum. The interest in the anticancer action of this particular carotenoid is relatively recent and can be explained by several reasons:
a. Among the common dietary carotenoids lycopene has the highest singlet oxygen quenching capacity and a high capability of quenching other free radicals in vitro.
b. The inverse relationship between lycopene intake or serum values and cancer risk that has been observed in particular for cancers of the prostate, pancreas, bladder and cervix.
c. Laboratory findings demonstrate that lycopene inhibits cancer cell growth in vivo and in vitro (in some cases independently of their role as antioxidants).
d. New evidence has provided a mechanistic explanation for the anticancer activity of lycopene.

Nutrition and absorption of lycopene.

Since mammals cannot synthesize carotenoids, including lycopene, these pigments are obtained mainly from vegetables and fruits. In contrast to α-carotene, β-carotene, lutein and zeaxanthin, which are widely distributed among a great variety of fruits and vegetables, lycopene enters the diet predominantly in tomatoes and tomato products. These include popular dishes such as chili con-carne, pizza or spaghetti with tomato sauce. Because of the frequency of consumption, tomato ketchup is also a major contributor of lycopene in many diets (1). The loss of lycopene during food preparation, such as cooking, is minimal. In fact, bioavailability of lycopene can be greatly improved by heating with the addition of some fat, as has been demonstrated by Stahl and Sies (2).

In Western countries lycopene is the major carotenoid in human plasma; its concentration (~0.8 μM) is more than twice that of β-carotene or the sum of lutein and zeaxantin together. This may result from either its high intake in the diet or its long

©1998 American Chemical Society

half-life in the body. The latter suggests that its frequent supply in the diet may not be contributory, but Micozzi (3) has stressed the importance of frequent intake of lycopene. Healthy subjects under strict carotenoid intake regimens were studied. Lycopene blood levels, but not those of α-and β-carotene or lutein, fell significantly during the 60 day follow-up when a carotenoid diet was not supplied. This suggests that plasma carotenoid concentrations, other than lycopene, remain fairly constant and are slow to change with low dietary intake. In another similar study (4), the fall in the plasma lycopene level was accompanied by a fall of other major carotenoids in the plasma. The reasons for the discrepancy between these two studies are not clear. Nevertheless, the two studies reported a steep fall in lycopene plasma levels in subjects with a low carotenoid diet.

Of the different geometrical isomers of lycopene, the cis isomers (9-and 13-cys) are better absorbed than the all-trans form (2). The absorption and distribution of [^{14}C]lycopene were studied in rats and in rhesus monkeys following the oral administration of the carotenoid in olive oil. The liver contained the largest amount of radioactive pigment. No labeled metabolic products of lycopene were found (5).

Physiological and pathological conditions which may affect lycopene serum levels.

The fact that age and stress significantly reduce lycopene serum levels indicates the appropriateness of specific populations for supplementation programs. A recent study was undertaken to characterize plasma concentrations of α-tocopherol, lycopene, β-carotene and several other carotenoids in elderly subjects. The test population consisted of 94 participants, 77 - 99 years old accrued from a project known as the Nun Study, (a longitudinal study of aging and Alzheimer disease patients). Concentrations of all analytes, except lycopene, were similar to or higher than those reported for several middle-aged American populations. Lycopene concentrations were significantly lower in the nun population as compared with the middle-aged populations and tended to decrease across age groups (6).

In another study, which tested age-dependent changes in antioxidant plasma levels in a high risk stomach cancer population (1,364 subjects 35-69 years of age), lycopene was the only antioxidant whose level decreased significantly (7).

Investigation was carried out to determine whether the acute phase response is associated with suppressed circulating levels of antioxidants in a population of 85 Catholic sisters 77-99 years of age (Nun Study). It was clearly demonstrated that such physiological conditions are associated with decreased plasma levels of the antioxidants lycopene, α-carotene, and β-carotene. The authors concluded that this decrease in circulating antioxidants may further compromise antioxidant status and increase oxidative stress and damage in the eldery (8).

Another recognized age-related pathology is macular degeneration. It is well documented that lutein and zeaxanthin are the main carotenoids in the eye macula. A study was designed to investigate the relationship between tocopherol levels and various carotenoids in the serum with age-related macular degeneration. Cases included 167 patients with various related pathologies and an equal number of controls. The surprising results show that the average levels of most carotenoids, including those composing macular pigment (lutein and zeaxanthin), were similar in cases and controls. However, persons with levels of lycopene in the lowest quintile were twice as likely to have age-related macular degeneration. The authors concluded that very low levels of lycopene, but not other dietary carotenoids or tocopherols, are related to age-related macular degeneration (9).

Lycopene as an antioxidant.

Reactive oxygen species occur in tissues and can damage DNA, proteins, carbohydrates and lipids. Lycopene exhibits the highest physical quenching rate constant with singlet oxygen (10). In a recent study (11), the ability of β-carotene to quench NOO· radicals was compared to that of lycopene. NOO· radicals are present in tobacco smoke and may cause cancer by reacting with various cell components. Cell staining with trypan blue was used to follow cell death. It was found that lycopene was at least three-fold more effective than β-carotene in preventing cell death by quenching of NOO· radicals. Smedman et al. (12) have reported that lycopene protects DNA damage in colon cells induced by 1-methyl 3-nitro-1-nitrosoguanidine (MNNG) and H_2O_2. The greatest protective effect was observed at a lycopene concentration approximating that found in blood of subjects consuming normal levels of tomatoes. The authors concluded that this effect cannot be explained simply by the antioxidant capacity of lycopene because the genotoxic effect of the alkylating agents, MNNG, was also inhibited.

Epidemiological studies involving lycopene and cancer.

Beta-carotene was extensively studied and implicated as a cancer preventive agent (see (13) for a review). However, intervention studies carried out with this carotenoid have revealed no beneficial effect (14,15) nor showed a negative effect (16,17). In the well-known Finnish study (16) almost 30,000 heavy smokers (all male) were followed for five to eight years in a randomized trial. An unexpectedly higher incidence of lung cancer was observed among men who received β-carotene than among those who did not. In another study (15) the efficacy of β-carotene in preventing colorectal adenoma (a precursor of invasive carcinoma) was tested in randomly assigned patients. From the results, it is clear that there is no evidence indicating that β-carotene reduces the incidence of adenomas. These results point out that dietary factors other than β-carotene may contribute to a reduction in cancer risk associated with a diet high in vegetable and fruits.

No intervention trials in humans investigating the potential effect of lycopene supplementation on the prevention of cancer have been performed to date. At present, only epidemiological studies which examined data on lycopene intake (based on questionnaires) or lycopene plasma levels in relation to cancer risk are available.

A study of cervical intra-epithelial neoplasia was designed to test the potential protective effects of various carotenoids (18). In this study, both dietary and serum lycopene manifested a strong inverse association with this malignancy. No such association was detected in the same study in regard to β-carotene intake. A lower level of serum lycopene was also observed in patients who subsequently developed bladder (19) and pancreatic (20) cancers. Serum levels of retinol, retinol-binding protein, and β-carotene were similar among cases and controls.

A protective effect of vitamin-A and β-carotene was found in squamous cell and small cell lung cancers. (21). In a later study, however, β-carotene-rich foods such as papaya, sweet potato, mango and yellow orange vegetables showed little influence on survival of lung cancer patients (22). The authors concluded that β-carotene intake before diagnosis of lung cancer does not affect the progression of the disease. In contrast, a tomato-rich diet which contributes only small amounts of β-carotene to the total carotenoid intake had a strong positive relationship with survival, particularly in women. In other studies however, lycopene intake was found to be unrelated to lung cancer risk(23,24).

A significant trend in risk reduction of gastric cancers by high tomato consumption was observed in a study which estimated dietary intake in low risk versus high risk areas in Italy (25). It is interesting that a similar regional impact on stomach cancer risk was found also in Japan (26). Out of several micronutrients including vitamins A, C and D and β-carotene in plasma, only lycopene was strongly-inversely associated with stomach cancer . A consistent pattern of protection for many sites of digestive tract cancer was associated with an increased intake of fresh tomatoes (27). Nevertheles, the possibility exists that other benefits of a Mediterranean life style get confused with high lycopene intake from a tomato rich diet.

Various case control studies have examined the risk factors for breast cancer which are unrelated to lycopene intake or serum levels (28,29). No direct epidemiological data are available on the potential role of lycopene consumption in skin cancer. However, Mercado et al. (30) demonstrated that lycopene, but not β-carotene, levels decrease significantly in the area of UV irradiated skin.

Recently, Giovannucci et al. (31) published a survey based on the U.S. Health Professional Follow-up Study.on prostatic cancer which included 48,000 people participatiants. This 7-year study showed that lycopene intake from tomato-based products is related to a low risk of prostate cancer. Consumption of other carotenoids (β-carotene, α-carotene, lutein, β-cryptoxantine) or retinol was not associated with the risk of prostate cancer.

In vivo and *in vitro* evidence for the inhibition of cancer by lycopene

The anticancer activity of carotenoids has been reviewed by Krinsky (32,33) and more recent reviews have specifically addressed the anticancer activity of lycopene (34,35).

The effect of several retinoids and carotenoids, including lycopene and β-carotene, was tested on rat C-6 glioma cells (36). All the retinoids and carotenoids which were dissolved in DMSO inhibited cellular growth at the 10 μM range. Lycopene effects were similar to those of β-carotene. Another study by Wang's group (37) was performed on an in vivo model of glioma cells transplanted in rats. This study also demonstrated that lycopene is an effective inhibitor. More recently, Nagasawa and colleagues (38) found that lycopene inhibits spontaneous mammary tumor development in SHN virgin mice probably by modulating the immune system in tumor-bearing mice (39).

The inhibitory effect of four carotenoids found in human blood and tissues which are effective against the formation of colonic aberrant crypt foci induced by N-methylnitrosourea was examined in Sprague-Dawley rats (40). Lycopene, lutein, α-carotene and palm carotenes (a mixture of alpha-carotene, beta-carotene and lycopene), but not beta-carotene, inhibited the development of aberrant crypt foci. The same research team also compared the cancer preventive effect of five kinds of carotenoids on mouse lung carcinogenesis with similar results (personal communication).

Our laboratory also tested lycopene, in comparison to β-carotene, in a well-known animal model for hormone-dependent mammary cancer dimethyl benz(a)anthracene (DMBA)-induced rat mammary tumors (41). The carotenoids (10 mg/kg, i.p.) were administered twice a week. The number of tumors was higher in the control (non-injected group) and in the β-carotene-treated group, than in the lycopene group. In the latter, the difference was evident only at longer periods of time. The largest average size of tumors was found in the β-carotene group and the smallest average size was in the lycopene group. This difference is statistically significant (p<0.05).

In another study, however, α-carotene was found to be more effective than β-carotene in inhibiting the proliferation of the human neuroblastoma cell line GOTO (42). Lycopene and other carotenoids were found to reduce hepatocytes cell injury induced by carbon tetrachloride. (43). Several carotenoids including lycopene protect against liver tumor promoter microcystin-LR (44).

We compared the potential anticancer activity of lycopene to that of α- and β-carotene in human cancer cells in culture. Lycopene was observed to strongly inhibit the growth of endometrial (Ishikawa) and lung (H226) cancer cells. The inhibition was dose-dependent - half maximal inhibition was at ~2 µM. α- and β-carotene were far less effective. For example, in Ishikawa endometrial cancer cells, a four-fold higher concentration of α-carotene or a ten-fold concentration of β-carotene were needed for growth inhibition of the same order. Inhibition of cell growth by lycopene was detected after 24 hours of incubation and was maintained for at least three days. In contrast to cancer cells, human fibroblasts were less sensitive to inhibition by lycopene, and the cells gradually escaped inhibition (45,46).

Attempts to elucidate the mechanism for the anticancer activity of lycopene.

Induction of cell to cell communication. The interesting observation that both β-carotene and canthaxanthin and to a lesser extend lycopene can inhibit malignant transformation caused by either methylcholanthrene - (MCA) or x-ray radiation in C3H/IOTI/2 cells has been reviewed (47). This study has been extended to other carotenoids. In addition to β-carotene and canthaxanthin, α-carotene and lycopene were also effective in inhibiting MCA-induced malignant transformation (48). In these studies, the effective dose (0.3 µM) of carotenoids, including lycopene, was lower than that found in previous studies of cell proliferation (10 µM). In a critical review, Wolf (49) interpreted the low concentration needed for carotenoid action (0.3 µM) to signify that it acts by catalytic or genomic (receptor) mechanisms, rather than as an antioxidant which may need higher concentrations. Support for this genomic action is found in a recent report by Bertram (50) which showed that carotenoids increase the expression of a gene encoding connexin43 - a gap junction protein. This effect was independent of provitamin-A or the antioxidant properties of carotenoids.

Inhibition of autocrine growth factor effects. One of the reasons for the uncontrolled growth of cancer cells is the augmented secretion of autocrine or paracrine growth factors which stimulate cell proliferation via different signaling pathways. The autocrine role of the IGF system in human mammary and endometrial cancer cells has been studied in our laboratory (51,52). We hypothesize that a possible mechanism for the inhibitory effect of lycopene may involve the modulation of the complex cellular IGF systems. IGF-I significantly stimulated [3H]thymidine incorporation, whereas lycopene inhibited both basal and IGF-I-induced cell proliferation. We found it impressive that lycopene is more effective in inhibiting fast growing cells than slow-growing ones. When 0.4 µM lycopene was applied, [3H]thymidine incorporation was inhibited only in cells growing in the presence of IGF-I (53). The same lycopene concentration was ineffective in cells growing slowly in the absence of IGF-I. Lycopene treatment did not affect the number or affinity of IGF-I receptors but significantly reduced IGF-induced tyrosine phosphorylation of insulin receptor substrate-1 and up-regulation of binding capacity of AP-1 transcription complex. These effects of the carotenoid may be explained by the accompanied increase in membrane-associated IGF-binding proteins. Recently we demonstrated that IGFBPs specifically associated with the cell membrane inhibits IGF-IR signaling in an IGF-dependent manner as revealed by measurement of short- and middle-term receptor mediated responses (54).

Intervention in cell cycle progression. We also performed experiments to analyze the effect of lycopene on the cell cycle in mammary lung and endometrial cancer cells (55). Lycopene slowed down cell cycle progression induced by IGF-I but did not affect the distribution of cells throughout the cell cycle phases in unsynchronized cells. In these cells, the inhibitory effect of lycopene was not accompanied by either necrotic or apoptotic cell death. As the carotenoid slowed down cell cycle progression in cells synchronized by mimosine and elevated cellular expression of p27 (the multi-cyclin dependent kinase inhibitor), interference by lycopene at several points of cell cycle progression is suggested. This lycopene-induced delay in S phase progression was observed in all cancer cell lines tested.

Synergism with other anticancer active compounds. Countryman et al. (56) reported that lycopene inhibits the growth of HL-60 cells, a well-known model for promyelocytic leukemia. These cells are also inhibited when *all-trans* retinoic acid is administered in micromolar concentration. We have recently confirmed this observation (unpublished data). One of the drawbacks of retinoid therapy is its high toxicity. To avoid this problem a combination therapy of several drugs was considered (57). Indeed in our study a hundred-fold lower concentration of retinoic acid, which does not inhibit cancer cell proliferation, induced a synergistic inhibitory effect when applied together with a low concentration of lycopene. The mechanism of this effect and the role of cell differentiation in this process is under investigation.

Summary

Prevention of tumor induction and inhibition of tumor growth by natural food constituents is an intriguing idea which has promoted numerous studies. Among various plant constituents, carotenoids have been extensively studied and implicated as cancer preventive agents. However, the idea of a single "magic bullet" which had been successful in *in vitro* and *in vivo* models failed to show beneficial effects in several clinical studies. Thus, our hypothesis is that anticancer effects of plant-derived constituents, such as carotenoids, relates to their ability to synergize with other anti-cancer compounds such as Vitamin-D_3 and retinoic-acid, which when used alone are active only at high and toxic concentrations. As lycopene levels may be low and vary considerably in individuals, the elucidation of a lycopene mechanism of action in inhibiting cancer cell growth will endorse its inclusion as an important component of dietary regimens.

Acknowledgments

The authors gratefully acknowledge the various preparations of tomato lycopene received from LycoRed, Natural Products Industries, Ltd. (Beer-Sheva, Israel). This work was supported in part by a grant (to M.D.) from the Chief Scientist, Israel Ministry of Health.

References

1. Mangels, A. R.; Holden, J. M.; Beecher, G. R.; Forman, M. R.; Lanza, E. *J. Am. Diet Assoc.* **1993** *93,* 384-296.
2. Stahl, W.; Sies, H. *J. Nutr.* **1992** *122,* 2161-6.
3. Micozzi, M. S.; Brown, E. D.; Edwards, B. K.; Bieri, J. G.; Taylor, P. R.; Khachik, F.; Beecher, G. R.; Smith, J. C. *Am. J. Clin. Nutr.* **1992** *55,* 1120-5.
4. Rock, C. L.; Swendseid, M. E.; Jacob, R. A.; Mckee, R. W. *J. Nutr.* **1992** *122,* 96-100.
5. Mathews-Roth, M. M.; Welankiwar, S.; Sehgal, P. K.; Lausen, N. C.; Russett, M.; Krinsky, N. I. *J. Nutr.* **1990** *120,* 1205-13.
6. Gross, M. D.; Snowdon, D. A. *Nutr Res* **1996** *16,* 1881-90.
7. Buiatti, E.; Munoz, N.; Kato, I.; Vivas, J.; Muggli, R.; Plummer, M.; Benz, M.; Franceschi, S.; Oliver, W. *Int J Cancer* **1996** *65,* 317-22.
8. Boosalis, M. G.; Snowdon, D. A.; Tully, C. L.; Gross, M. D. *Nutrition* **1996** *12,* 475-8.

9. Mares Perlman, J. A.; Brady, W. E.; Klein, R.; Klein, B. E.; Bowen, P.; Stacewicz, S. M.; Palta, M. *Arch Ophthalmol* **1995** *113*, 1518-23.
10. Di Mascio, P.; Kaiser, S.; Sies, H. *Arch. Biochem. Biophys.* **1989** *274*, 532-8.
11. Bohm, F.; Tinkler, J. H.; Truscott, T. G. *Nature Medicine* **1995** *1*, 98-9.
12. Smedman, A. E. M.; Smith, C.; Davison, I. R.; Rowland, I. R. *Anticancer Res. Abs 82* **1995** *15*, 1656.
13. Van Poppel, G. *Eur. J. Cancer* **1993** *29A*, 1335-44.
14. Hennekens, C. H.; Buring, J. E.; Manson, J. E.; et al. *New Engl. J. Med.* **1996** *334*, 1145-9.
15. Greenberg, E. R.; Baron, J. A.; Tosteson, T. D.; et al. *New Engl. J. Med.* **1994** *331*, 141-7.
16. Heinonen, O. P.; Huttunen, J. K.; Albanes, D.; et al. *New Engl. J. Med.* **1994** *330*, 1029-35.
17. Omenn, G. S.; Goodman, G. E.; Thornquist, M. D.; et al. *New Engl. J. Med.* **1996** *334*, 1150-5.
18. VanEenwyk, J.; Davis, F. G.; Bowen, P. E. *Int. J. Cancer* **1991** *48*, 34-9.
19. Helzlsouer, K. J.; Comstock, G. W.; Morris, J. S. *Cancer Res* **1989** *49*, 6144-8.
20. Burney, P. G.; Comstock, G. W.; Morris, J. S. *Am. J. Clin. Nutr.* **1989** *49*, 895-900.
21. Ziegler, R. G.; Mason, T. J.; Stemhagen, A. *Am. J. Epidemiol* **1986** *123*, 1080-93.
22. Goodman, M. T.; Kolonel, L. N.; Wilkens, L. R.; Yoshizawa, C. N.; Lemarchand, L.; Hankin, J. H. *Eur. J. Cancer* **1992** *28*, 495-501.
23. Le Marchand, L.; Yoshizawa, C. N.; Kolonel, L. N.; Hankin, J. H.; Goodman, M. T. *J Natl Cancer Inst* **1989** *81*, 1158-64.
24. Steinmetz, K. A.; Potter, J. D.; Folsom, A. R. *Cancer Res.* **1993** *53*, 536-43.
25. Buiatti, E.; Palli, D.; Decarli, A.; et al. *Int J Cancer* **1990** *45*, 896-901.
26. Tsugane, S.; Tsuda, M.; Gey, F.; Watanabe, S. *Environ Health Perspect* **1992** *98*, 207-10.
27. Franceschi, S.; Bidoli, E.; La Vecchia, C.; Talamini, R.; D'Avanzo, B.; Negri, E. *Int J Cancer* **1994** *59*, 181-4.
28. London, S. J.; Stein, E. A.; Henderson, I. C.; Stampfer, M. J.; Wood, W. C.; Remine, S.; Dmochowski, J. R.; Robert, N. J.; Willett, W. C. *Cancer Causes Control* **1992** *3*, 503-12.
29. Potischman, N.; McCulloch, C. E.; Byers, T.; et al. *Am J Clin Nutr* **1990** *52*, 909-15.
30. Ribaya Mercado, J. D.; Garmyn, M.; Gilchrest, B. A.; Russell, R. M. *J Nutr* **1995** *125*, 1854-9.
31. Giovannucci, W.; Ascherio, A.; Rimm, E. B.; Stampfer, M. J.; Colditz, G. A.; Willett, W. C. *J Natl Canc Inst* **1995** *87*, 1767-76.
32. Krinsky, N. I. *Am. J. Clin. Nutr.* **1991** *53*, 238S-46S.
33. Krinsky, N. I. in *Free Radicals and Aging;* (Emers, I., Chance, B., Eds) Verlag: Basel, 1992, 227-34.
34. Gerster, H. *J Am Coll Nutr* **1997** *16*, 109-26.
35. Stahl, W.; Sies, H. *Arch Biochem Biophys* **1996** *336*, 1-9.
36. Wang, C.-J.; Lin, J.-K. *Proc. Natl. Sci. Counc. B. ROC* **1989** *13*, 176-83.
37. Wang, C. J.; Chou, M. Y.; Lin, J. K. *Cancer Letters* **1989** *48*, 135-42.
38. Nagasawa, H.; Mitamura, T.; Sakamoto, S.; Yamamoto, K. *Anticancer Res.* **1995** *15*, 1173-8.
39. Kobayashi, T.; Itjima, K.; Mitamura, T.; Torilzuka, K.; Cyong, J.; Nagasawa, H. *Anticancer drugs* **1996** *7*, 195-8.
40. Narisawa, T.; Fukaura, Y.; Hasebe, M.; Ito, M.; Aizawa, R.; Murakoshi, M.; Uemura, S.; Khachik, F.; Nishino, H. *Cancer Lett* **1996** *107*, 137-42.
41. Sharoni, Y.; Giron, E.; Rise, M.; Levy, J. *Cancer Detect Prevent* **1997** *21*, 118-23.

42. Murakoshi, M.; Takayasu, J.; Kimura, O.; et al. *J. Natl. Canc. Inst.* **1989** *81*, 1649-52.
43. Kim, H. *Int J Biochem Cell Biol* **1995** *27*, 1303-9.
44. Matsushima-Nishiwaki, R.; Shidoji, Y.; Nishiwaki, S.; Yamada, T.; Moriwaki, H.; Muto, Y. *Lipids* **1995** *30*, 1029-34.
45. Levy, J.; Bosin, E.; Feldman, B.; Giat, Y.; Miinster, A.; Danilenko, M.; Sharoni, Y. *Nutr. & Cancer* **1995** *24*, 257-67.
46. Sharoni, Y.; Levy, J. in *Proceedings of the XVI international cancer congress;* (Rao, R. S., Deo, M. G., Sanghvi, L. D., Eds) Monduzzi Editore: Bologna, 1994, Vol. 1; pp 641-5.
47. Bertram, J. S.; Pung, A.; Churley, M.; Kappock, T. d.; Wilkins, L. R.; Cooney, R. V. *Carcinogenesis* **1991** *12*, 671-8.
48. Zhang, L.-X.; Cooney, R. V.; Bertram, J. *Carcinogenesis* **1991** *12*, 2109-14.
49. Wolf, G. *Nutr. Rev.* **1992** *50*, 270-4.
50. Zhang, L. X.; Cooney, R. V.; Bertram, J. S. *Cancer Res.* **1992** *52*, 5707-12.
51. Kleinman, D.; Roberts Jr., C. T.; LeRoith, D.; Schally, A. V.; Levy, J.; Sharoni, Y. *Regulatory Peptides* **1993** *48*, 91-8.
52. Hershkovitz, E.; Marbach, M.; Bosin, E.; Levy, J.; Roberts Jr., C. T.; LeRoith, D.; Schally, A. V.; Sharoni, Y. *J. Clin. Endo. Metab.* **1993** *77*, 963-8.
53. Sharoni, Y.; Levy, J. in *Natural antioxidants and food quality in atherosclerosis and cancer prevention;* (Kumpulainen, J. T., Salonen, J. K., Eds) The Royal Society of Chemistry.: Cambridge, 1996, 378-85.
54. Karas, M.; Danilenko, M.; Fishman, D.; LeRoith, D.; Levy, J.; Sharoni, Y. *J Biol Chem* **1997** *272*, 16514-20.
55. Levy, J.; Karas, M.; Amir, H.; Giat, J.; Danilenko, M.; Sharoni, Y. *Anticancer Res. Abs 80* **1995** *15*, 1655.
56. Countryman, C.; Bankson, D.; Collins, S.; Man, B.; Lin, W. *Clin. Chem.* **1991** *37*, 1056.
57. Chen, A.; Licht, J. D.; Wu, Y.; Hellinger, N.; Scher, W.; Waxman, S. *Blood* **1994** *84*, 2122-9.

NATURAL PLANTS

Fruits and Related Compounds

Chapter 5

Mechanisms of Phytochemical Inhibition of Carcinogenesis: Elucidating the Role of γ-Tocopherol in Nutrition

T. S. Burnett, Y. Tanaka, P. J. Harwood, and R. V. Cooney

University of Hawaii Cancer Research Center, 1236 Lauhala Street, Honolulu, HI 96813

Epidemiologic studies have consistently demonstrated a protective effect for fruit and vegetable consumption against the development of many forms of cancer and heart disease. Identifying the chemical constituents responsible for observed health benefits and their mechanism(s) of action are critical to optimizing public health. A number of recent studies have found that serum γ-tocopherol levels are inversely related to risk for cardiovascular disease and some cancers, despite the fact that sources of γ-tocopherol are limited to dietary oils of plant origin that contain significant quantities of fats. In particular, tocopherols are often found to be highly concentrated in the germ plasm of seeds and we observe the selective concentration of γ-tocopherol in the germ of peanuts. Whereas α-tocopherol is considered to be the most biopotent of the vitamin E analogues, γ-tocopherol is observed to be more effective in preventing neoplastic transformation and cellular damage induced by cytokines. Studies indicate that the mechanism of action for γ-tocopherol in preventing cellular damage may be unique relative to other phytochemicals, and potentially related to specific chemical and biological properties of γ-tocopherol, including the ability to chemically reduce NO_2 to NO, enhance cellular NO synthesis, alter the kinetics of cell growth, enhance cell saturation density, and reduce DNA strand breaks in C3H 10T1/2 murine fibroblasts.

Vitamin E is a lipid soluble vitamin initially recognized as a dietary factor essential for reproduction in 1922 by Evans and Bishop (reviewed in 1). It has subsequently been characterized and found to consist of a family of closely related chromanols (α-, β-, δ- γ-tocopherols and tocotrienols) that contribute in varying extents to vitamin E function in animals. Although over 70 years have passed since the discovery of vitamin E, the mechanism(s) accounting for its bioactivity remains elusive. Localized in cell membranes, vitamin E is a major free radical chain-breaking antioxidant, and has been

46

reported to modulate signal transduction pathways regulating cellular proliferation. It is unclear, however, whether its function as a vitamin is related to these properties. Clinical deficiency in animals has been associated with fetal death and resorption. Deficiency of vitamin E in humans, most commonly associated with malabsorption syndromes, can result in progressive neurological syndromes, skeletal muscle dysfunction, and erythrocyte fragility, and is thought to be associated with increased risk for cardiovascular and degenerative age-related diseases and cancer.

Tocopherol Function: Relation to Vitamin E

Tocopherols are found ubiquitously in plants, particularly in association with seed oils, where they function to suppress oxidation of unsaturated fatty acids (2,3). Chloroplasts contain primarily α-tocopherol, whereas in seed oils, γ-tocopherol often predominates (4). The reasons for the differential accumulation of specific tocopherols in different plant compartments is not established, however, it may be related to functional chemical differences in the reactions that occur between oxidants and the tocopherols in light versus dark conditions. In corn the concentration of tocopherol in the embryo-containing germ is greater than that found in the surrounding endosperm (5). Wheat germ also concentrates tocopherols with α- and β-tocopherol as the predominant forms (4). As shown in figure 1, we observe a selective concentration of γ-tocopherol within the germ of peanuts relative to the endosperm, suggesting a possible role for γ-tocopherol in protecting the dormant embryo.

γ-Tocopherol comprises 70% of the tocopherols consumed in the U.S. diet (6) and is derived from many plant sources including nuts, grains, and seeds, as well as the food products derived from them (4). In contrast α-tocopherol, the other major naturally-occurring tocopherol, constitutes approximately 80% of the tocopherols found in human plasma and is significantly more bioactive as a source of vitamin E activity (7). As a consequence, most research has focused on the role of α-tocopherol in nutrition and the prevention of chronic disease to the exclusion of other tocopherols. Recently a number of studies in diverse research areas have raised significant questions regarding the assumption that γ-tocopherol is simply a less effective form of vitamin E without any intrinsic nutritional role. Yamashita et al. (8) reported that γ-tocopherol administered in whole sesame seeds, or when given in combination with sesame seed lignans, was equally biopotent with α-tocopherol and could completely fulfill the vitamin E requirement of rats with a concomitant rise in plasma levels of γ-tocopherol. Further research has demonstrated that sesame lignans increase tissue concentrations of γ-tocopherol, even in the presence of adequate α-tocopherol levels in the diet and plasma (9). These studies raise significant questions relating to the practice of calculating vitamin E equivalents for epidemiological studies in which dietary intake or plasma levels of γ-tocopherol are multiplied by its fractional biopotency, based on rodent studies using purified tocopherols. In the case of plasma levels it could be argued that γ-tocopherol and α-tocopherol have equal biopotencies, whereas the dietary source of γ-tocopherol may play an important role in determining subsequent plasma levels and corresponding bioactivity.

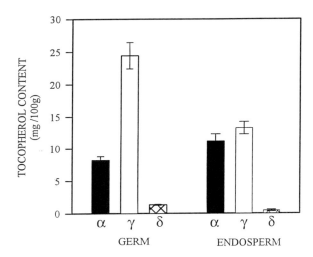

Figure 1. Comparison of tocopherol levels in the germ and endosperm of peanuts. Lyophilized peanut (raw Virginia fancy peanuts provided by Hoody's Corp., Los Angeles, CA) germ and endosperm were separated, crushed, and extracted repeatedly with ethanol containing BHT (50 μg/ml) and heated to 60°C for 30 minutes. The ethanol extracts were combined in a separatory funnel, an equal volume of aqueous 1.34% NaCl was added, and the sample extracted 3x with hexane. The hexane extracts were combined, washed with H_2O, and evaporated. The sample was then reconstituted in 1ml of HPLC mobile phase (hexane:isopropanol, 99:1) and chromatographed on a Phenomenex Lichrosorb normal phase column (5μM silica, 60A, 250 mm x 4.6 mm) at a flow rate of 1 ml/min (UV monitoring at 295 nm). Data plotted represent the mean ± SEM for five separate analyses.

48

Recently a metabolite of γ-tocopherol, 2,7,8-trimethyl-2-(β-carboxyethyl)-6-hydroxychroman, has been identified as a natriuretic factor in humans, responsible for sodium excretion and controlling the pool of extracellular fluid in the body, whereas the corresponding compound derived from α-tocopherol was found to be ineffective in promoting natriuresis (10). In light of these recent observations and our incomplete understanding of the mechanism for vitamin E function, the need for additional basic research is clear.

Associations between Vitamin E and Disease

Traditionally, epidemiologic investigations have focused either entirely on α-tocopherol in the diet and/or plasma or relied on estimates of vitamin E equivalents, which tend to discount the potential contribution from γ-tocopherol, as the biopotency of γ-tocopherol is estimated to be 10-20% that of α-tocopherol on the basis of rat bioassays (7). While it is not known if human uptake and retention of γ-tocopherol is affected by dietary source, as demonstrated in rats (8), it is reasonable to speculate that plasma and tissue levels may be influenced analogously when γ-tocopherol is derived from unrefined foods such as nuts and seeds, as opposed to highly refined cooking oils. In this regard, it is of interest that a number of recent studies have shown protective effects against cardiovascular disease incidence and mortality by consumption of nuts, which represent a major unrefined food source for γ-tocopherol (4). Fraser et al. (11) reported a significant decrease in the risk of myocardial infarction and cardiac death associated with nut consumption in a prospective study of California Seventh-Day Adventists, while Ohrvall et al. (12) found reduced serum levels of γ-tocopherol in patients with cardiovascular disease relative to controls. In a concomitant cross-sectional study of Swedish and Lithuanian men, Kristenson et al. (13) reported that lycopene and γ-tocopherol levels were nearly twice as high in the Swedish population, while Swedish men were four times less likely to die of cardiovascular disease. However, no differences in α-tocopherol or cholesterol levels were observed between the two study populations. Prineas et al. (14) also reported a dose-dependent decrease in risk of death from coronary heart disease associated with dietary nut intake. Consumption of walnuts, an excellent source of γ-tocopherol, was found to significantly lower serum cholesterol and improve lipid profiles (15), despite containing >80% of calories in the form of fat. Kushi et al. (16) reported that dietary vitamin E consumption was strongly inversely correlated with death from coronary heart disease in a prospective cohort study of post-menopausal women, however, consumption of vitamin E supplements (containing exclusively α-tocopherol) was unrelated to mortality risk. On the other hand, recent intervention studies have shown strong protective effects for α-tocopherol supplementation in the prevention of heart disease (17,18) and Alzheimer's disease (19), although other intervention studies have shown no net benefit to overall mortality from α-tocopherol supplementation (20).

Epidemiologic evidence for an association between dietary γ-tocopherol intake and cancer incidence has only recently begun to emerge. Significant inverse associations between serum γ-tocopherol level and cancer incidence have been observed for early stage cervical cancer (21) and in an ecologic study of lung cancer incidence comparing Fijians with Cook Islanders (22,23). In a cohort study of Japanese Americans, Nomura et al. (24) observed a significant inverse association with oral/pharyngeal and laryngeal tumors for γ-tocopherol, but found no association with esophogeal tumors. Zheng et al. (25) however, showed a significant positive association for serum selenium and γ-tocopherol levels with oral and pharyngeal tumors. The strong correlation between γ-tocopherol and dietary polyunsaturated fat may confound observations of the effects of γ-tocopherol, as γ-tocopherol in the diet is highly correlated with fatty acid unsaturation, particularly non-marine polyunsaturated fatty acids which may significantly enhance oxidative damage (26).

Chemical and Biological Effects of Tocopherols

Biological and chemical differences have also been shown for the two main tocopherol analogues. γ-Tocopherol is taken up more rapidly by human endothelial cells compared to α-tocopherol (27), is superior to α-tocopherol in preventing neoplastic transformation in C3H 10T1/2 fibroblasts (28), and is better able to protect against DNA single-strand breaks caused by cellular exposure to NO_2 (29), as well as cell death associated with the endogenous generation of nitric oxide (30). Recently, Christen et al. (31) demonstrated the superiority of γ-tocopherol over α-tocopherol in preventing peroxynitrite-mediated lipid hydroperoxide formation and suggested that γ-tocopherol may fill a unique biological role by reacting with potentially damaging electrophiles that are generated endogenously. This conclusion is based, in part, upon the unique chemical characteristics of γ-tocopherol, in which the absence of a methyl group at the C-5 position of the chromanol ring creates a reactive nucleophilic center which can scavenge electrophiles, thereby inhibiting oxidative modification of key cellular molecules. γ-Tocopherol has also been shown to react with the NO_2 radical at the C-5 position to form nitric oxide and tocored (the oxidized quinone of γ-tocopherol), or to form 5-nitro-γ-tocopherol when the free radical is localized on the nitrogen atom of NO_2 (30). In contrast, α-tocopherol reacts with NO_2 to form an intermediate capable of nitrosating amines (28) and does not generate NO efficiently from NO_2 in the absence of light or heat (30). The superiority of γ-tocopherol in reducing the highly reactive NO_2 radical to the more stable NO radical may be an important factor in the ability of γ-tocopherol to reduce NO_2-mediated DNA single-strand breaks (29). While the endogenous generation of NO has been linked to DNA damage (32), carcinogenesis (33,34), and other pathologies (35) resulting from its oxidation to more reactive species (32,36) such as peroxynitrite and NO_2, NO in and of itself possesses considerable antioxidant character (37). Furthermore, NO may have important cell signaling functions (38) that mitigate any indirect harmful effects associated with its oxidation products. In this regard we observe that γ-tocopherol, but not α-tocopherol, enhances cytokine-induced production of NO by C3H 10T1/2 cells in culture (Figure 2), as measured by the stable end product, nitrite. Despite the increased production of NO in these cells treated with interferon-γ (IFNγ) and bacterial lipopolysacharride (LPS), which is normally associated with increased DNA single-strand breaks, γ-tocopherol reduced the level of breaks below that of cytokine-treated control cells (Figure 3).

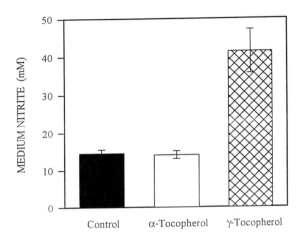

Figure 2. Enhanced NO generation in cytokine-stimulated C3H 10T1/2 cells by γ-tocopherol. C3H 10T1/2 cells were grown to confluence then treated with 10 ng/ml interferon-γ (IFNγ) and 10 μg/ml bacterial lipopolysaccharide (LPS) at the time of media change. At the same time, cultures were treated with either α-tocopherol (30 μM) or γ-tocopherol (30 μM) in ethanol (25 μl/5ml media), or 25 μl ethanol as control. NO production was measured as mean nitrite ± SEM in the medium 72 hours later (n = 3).

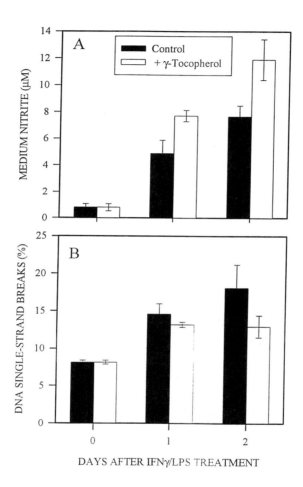

Figure 3. Effect of γ-tocopherol on DNA single-strand breaks in C3H 10T1/2 cells induced to generate NO. Nitrite levels in the media (Panel A) were determined for cells grown and treated with IFNγ/LPS as described in Figure 2 with or without 10 μM γ-tocopherol. DNA single-strand breaks (Panel B) were determined according to the method of Bittrich et al. (*29*) and data is presented as the % DNA eluted relative to that retained by the filter. Each value represents the mean of two samples. Error bars indicate the high and low values.

Although the potential for oxidative damage from cellular generation of nitrogen oxides is well established, the role of NO in the carcinogenic process is unclear. Chemical inhibitors of NO synthesis block the formation of neoplastic foci in the standard C3H 10T1/2 transformation assay (33), whereas treatment with antisense deoxyoligonucleotides, which specifically reduces levels of both NO synthase mRNA and protein, causes enhanced foci formation (39). While further research is needed to resolve this apparent conflict, the possibility that under specific conditions enhanced NO formation may in fact protect against tumor development is consistent with the enhanced NO formation and chemopreventive effects observed for γ-tocopherol. The relative ratio of cellular superoxide anion and NO may play a significant role with respect to the extent and nature of oxidative damage that occurs in cells (40). Consequently, altering cellular synthesis of NO may be either beneficial or detrimental depending upon the concentrations of other key reductants and/or oxidants.

Effects of Tocopherols on Cellular Proliferation. In the C3H 10T1/2 transformation assay, agents which generally inhibit proliferation may reduce the number of neoplastic foci by inhibiting growth of neoplastic cells, without toxicity to quiescent normal cells. To test the effects of α- and γ-tocopherols on the growth of malignant cells, growth curves were determined in the presence of α-tocopherol, γ-tocopherol, and solvent control for the neoplastic MCA cell line, derived from malignant colonies of carcinogen-treated C3H 10T1/2 cells (Figure 4). Both tocopherols significantly elevated the growth rates of transformed cells above control cultures. From this experiment, we conclude that the tocopherols do not inhibit transformation through general growth inhibitory effects, rather they appear to stimulate growth of neoplastic cells. As a consequence, it is conceivable that high doses of tocopherol could accelerate the growth of a pre-existing tumor, while still acting as cancer preventive agents in normal cells by blocking mutations that could lead to neoplastic transformation. Such an explanation would be consistent with the study of Potischman et al. (21), who reported a protective association for γ-tocopherol with early stage cervical cancer, but found elevated γ-tocopherol levels in patients with advanced disease relative to controls.

We also examined the effects of tocopherols on the growth and saturation density of normal C3H 10T1/2 cells. In contrast to tumor cells, α-tocopherol had no significant effect on log-phase growth and slightly enhanced saturation density as shown in Figure 5. γ-Tocopherol, however, significantly decreased the growth rate between 72-96 hours after treatment (Figure 6) and also increased both saturation density and confluent density. The 72-96 hour time period corresponds to the time when maximum uptake of tocopherol occurs (data not shown). Growth rates subsequent to this time period were similar to control cells, despite the fact that media and cellular tocopherol levels remained high.

In confluent C3H 10T1/2 cells, growth factor-mediated DNA synthesis was inhibited approximately 50% by α-tocopherol and 25% by γ-tocopherol as compared to controls (Table I). Previously, Mordan et al. (38) showed that inhibition of DNA synthesis in confluent C3H 10T1/2 fibroblasts by retinoids may be an important element in their ability to prevent neoplastic transformation during the promotional

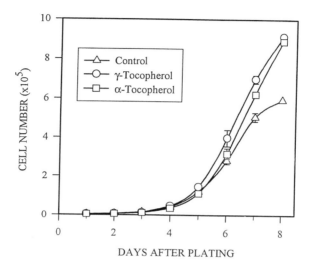

Figure 4. Effects of tocopherols on the growth of the transformed MCA cell line. MCA cells, a malignant cell line derived from cloning transformed C3H 10T1/2 cells obtained from a neoplastic focus, were plated at a density of 5000/35 mm dish on day 0, and treated with either α-tocopherol (30 μM), γ-tocopherol (30 μM), or solvent control (10 μl acetone/2.5 ml media) on day 1. Cell counts were obtained with a Coulter counter daily for 8 days and values represent the mean ± SEM for 4 culture dishes.

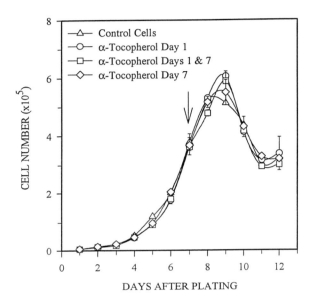

Figure 5. Growth curve of α-tocopherol-treated cultures. Cells were plated at a density of 5000/35 mm dish on day 0, and treated on day 1, days 1 and 7, or day 7 with α-tocopherol (30 μM). Control cells were treated with 10 μl of solvent vehicle (acetone). Cell counts were obtained daily for 12 days with a Coulter counter and each value represents the mean ± SEM for 4 culture dishes. Media was changed on day 7 and is designated by an arrow.

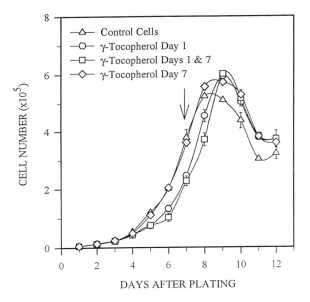

Figure 6. Growth curve of γ-tocopherol-treated cultures. Cells were plated at a density of 5000/35 mm dish on day 0, and treated on day 1, days 1 and 7, or day 7 with γ-tocopherol (30 μM as described in Figure 5). Cell counts were obtained daily and each value represents the mean ± SEM for 4 culture dishes. Media was changed on day 7 and is designated by an arrow.

phase. The results presented here suggest that while this effect may contribute to the ability of tocopherols to inhibit transformation, it does not explain the greater efficacy of γ-tocopherol relative to α-tocopherol in preventing foci formation, as γ-tocopherol has a lesser effect on proliferation at this stage. Furthermore, unlike the retinoids which cause decreased saturation density at growth arrest (*41*), both α- and γ-tocopherol increase cell saturation density at confluence (Figures 5 and 6).

Table I. Effects of α- or γ-Tocopherol on Growth Factor-Mediated DNA Synthesis in Confluent Cultures of C3H 10T1/2 Fibroblasts. Confluent C3H 10T1/2 cells in 96 well trays were incubated with either 30 μM α-, or γ-tocopherol, or vehicle control (0.5% acetone). Four days later, fresh cell culture medium was added and 11 hours after the media change, cultures were exposed to BrdU labeling agent. DNA synthesis was determined by measurement of BrdU incorporation using a cell proliferation assay kit #RPN 210 from Amersham Int., UK. Relative levels of BrdU incorporation based on absorbance at 410 nm are reported. Negative controls did not receive a media change at day 4 while positive controls did. Values represent the mean ± SEM. (n = 8).

Negative Control	Positive Control	α-Tocopherol	γ-Tocopherol
0	$0.026 \pm .001$	$0.014 \pm .002$	$0.020 \pm .002$

Conclusion

γ-Tocopherol represents a ubiquitous dietary molecule, whose potential functional significance has only recently begun to be appreciated. Its inverse association with risk and mortality for cardiovascular disease and cancer suggest either an intrinsic chemopreventive role or that γ-tocopherol may serve as a marker for other biologically active phytochemicals, as yet unidentified. The unique chemical and biological effects of γ-tocopherol that have been described support the former and suggest that this tocopherol may be an essential dietary component in the prevention of aging-related chronic disease. The effects of γ-tocopherol on cellular proliferation, mutation, antioxidant defense, and NO generation will hopefully provide clues that will enable future studies to identify the specific mechanisms responsible for the bioactivity of the tocopherols and determine the optimal dietary level of this and related vitamin E species.

Acknowledgments

This research was supported by Grant #CN-158 from the American Cancer Society (RVC).

Literature Cited

1. Machlin, L.J.; Brin, M.   Vitamin E.   In *Human Nutrition: A Comprehensive Treatise*; Alfin-Slater, R.B.; Kritchevsky, D., Eds.; Plenum Press: New York, NY, **1980**; Vol. 3B.
2. Yoshida, H.; Kajimoto, G.; Emura, S. *J. Am. Oil Chem. Soc.*, **1993**, *70*, 989-995.
3. Horwitt, M.K. *Vitam. Horm.*, **1962**, *20*, 541-544.
4. McLaughlin, P.J.; Weihrauch, J.L. *J. Am. Diet. Assoc.* **1979**, *75*, 647-665.
5. Grams, G.W.; Blessin, C.W.; Inglett, G.E. *J. Agric. Food. Chem.*, **1970**, *639*, 337-339.
6. Bieri, J.G.; Evarts, R.P. *J. Am. Dietetic Assoc.*, **1973**, *62*, 147-151.
7. Bieri, J.G.; Evarts, R.P. *J. Nutr.*, **1974**, *104*, 850-857.
8. Yamashita, K.; Nohara, Y.; Katayama, K.; Namiki, M. *J. Nutr.*, **1992**, *122*, 2440-2446.
9. Kamal-Elden, A.; Pettersson, D.; Appelqvist, L.-A. *Lipids*, **1995**, *30*, 499-505.
10. Wechter, W.J.; Kantoci, D.; Murray, Jr., E.D.; D'Amico, D.C.; Jung, M.E.; Wang, W.-H. *Proc. Natl. Acad. Sci. USA*, **1996**, *93*, 6002-6007.
11. Fraser, G.E.; Sabate, J.; Beeson, W.L.; Strahan, T.M. *Arch. Intern. Med.*, **1992**, *152*, 1416-1424.
12. Öhrvall, M.; Sundlöf, G; Vessby, B. *J. Intl. Med.*, **1996**, *239*, 111-117.
13. Kristenson, M.; Ziedén, B.; Kucinskienë, Z.; Elinder, S.L.; Bergdahl, B.; Elwing, B.; Abaravicius, A.; Razinkovienë, L.; Calkauskas, H.; Olsson, A.G. *Brit. Med. J.*, **1997**, *314*, 629-633.
14. Prineas, R.J.; Kushi, L.H.; Folsom, A.R.; Bostick, R.M.; Wu, Y. *N. Engl. J. Med.*, **1993**, *329*, 359.
15. Sabate, J.; Fraser, G.E.; Burke, K.; Knutsen, S.F.; Bennett, H.; Linsted, K.D. *N. Engl. J. Med.*, **1993**, *328*, 603-607.
16. Kushi, L.H.; Folson, A.R.; Prineas, R.J.; Mink, P.J.; Wu, Y.; Bostick, R.M. *N. Engl. J. Med.*, **1996**, *334*, 1156-1162.
17. Stampfer, M.J.; Hennekens, C.H.; Manson, J.E.; Colditz, G.A.; Rosner, B.; Willett, W.C. *N. Engl. J. Med.*, **1993**, *328*, 1444-1449.
18. Rimm, E.B.; Stampfer, M.J.; Ascherio, A.; Giovannucci, E.; Colditz, G.A.; Willett, W. *N. Engl. J. Med.*, **1993**, *328*, 1450-1456.
19. Sano, M.; Ernesto, C.; Thomas, R.G.; Klauber, M.R.; Schafer, K.; Grundman, M.; Woodbury, P.; Growdon, J.; Cotman, C.W.; Pfeiffer, E.; Schneider, L.S.; Thal, L.J. *N. Engl. J. Med.*, **1997**, *336*, 1216-1222.
20. The Alpha-Tocopherol, Beta-Carotene Cancer Prevention Study Group. *N. Engl. J. Med.*, **1994**, *330*, 1029-1035.
21. Potischman, N.; Herrero, R.; Brinton, L.A.; Reeves, W.C.; Stacewicz-Sapuntzakis, M.; Jones, C.J.; Brenes, M.M.; Tenorio, F.; de Britton, R.C.; Gaitan, E. *Am. J. Epi.*, **1991**, *134*, 1347-1355.
22. Kolonel, L.N.; Le Marchand, L.; Hankin, J.H.; Henderson, B.E. *Proc. Am. Assoc. Cancer Res.*, **1991**, *32*, 472.
23. Henderson, B.E.; Kolonel, L.N.; Dworsky, R.; Kerford, D.; Mori, E.; Singh, K.; Thevenot, H. *Natl. Cancer Inst. Monogr.*, **1985**, *69*, 73-81.

24. Nomura, A.M.Y.; Ziegler, R.G.; Stemmermann, G.N.; Chyou, P.-O.; Craft, N.E. *Cancer Epidemiol. Biomarker. Prev.*, **1997**, *6*, 407-412.
25. Zheng, W.; Blot, W.J.; Diamond, E.L.; Norkus, E.P.; Pate, V.; Morris, J.S.; Comstock, G.W. *Cancer Res.*, **1993**, *53*, 795-798.
26. Franke, A.A.; Harwood, P.J.; Shimamoto, T.; Lumeng, S.; Zhang, L.-X.; Bertram, J.S.; Wilkens, L.R.; Le Marchand, L.; Cooney, R.V. *Cancer Lett.*, **1994**, *74*, 17-26.
27. Tran, K.; Chan, A.C. *Lipids*, **1992**, *27,* 38-41.
28. Cooney, R.V.; Franke, A.A.; Harwood, P.J.; Hatch-Pigott, V.; Custer, L.J.; Mordan, L.J. *Proc. Natl. Acad. Sci. USA*, **1993**, *90*, 1771-1775.
29. Bittrich, H.; Mátzig, A.K.; Kráker, I.; Appel, K.E. *Chem. Biol. Interactions*, **1993**, *86*, 199-211.
30. Cooney, R.V.; Harwood, P.J.; Franke, A.A.; Narala, K.; Sundström, A.-K.; Berggren, P.-O.; Mordan, L.J. *Free Rad. Biol. Med.*, **1995**, *19*, 259-269.
31. Christen, S.; Woodall, A.A.; Shigenaga, M.K.; Southwell-Keely, P.T.; Duncan, M.W.; Ames, B.N. *Proc. Natl. Acad. Sci. USA,* **1997**, *94*, 3217-3222.
32. Wink, D.A.; Kasprzak, K.S.; Maragos, C.M.; Elespuru, R.K.; Misra, M.; Dunams, T.M.; Cebula, T.A.; Koch, W.H.; Andrews, A.W.; Allens, J.S.; Keefer, L.K. *Science*, **1991**, *254*, 1001-1003.
33. Mordan, L.J.; Burnett, T.S.; Zhang, L.-X.; Tom, J.; Cooney, R.V. *Carcinogenesis*, **1993**, *14*, 1555-1559.
34. Ohshima, H.; Bartsch, H. *Mutat. Res.*, **1994**, *305*, 253-264.
35. Miesel, R.; Kurpisz, M.; Kroger, H. *Free Radical Biol. & Med.,* **1996**, *20*, 75-81.
36. Arroyo, P.L., Hatch-Pigott, V.; Mower, H.F.; Cooney, R.V. *Mutat. Res.*, **1992**, *281*, 193-202.
37. Wink, D.A.; Hanbauer, I.; Krishna, M.C.; DeGraff, W.; Gamson, J.; Mitchell, J.B. *Proc. Natl. Acad. Sci. USA*, **1993**, *90*, 9813-9817.
38. Staeuble, B.; Boscoboinik, D.; Tasinato, A.; Azzi, A. *Eur. J. Biochem.*, **1994**, *226*, 393-402.
39. Lesoon-Wood, L.A.; Pierce, L.M.; Motosue, A.; Lau, A.F.; Cooney, R.V. *Proc. Am. Assoc. Cancer Res.*, **1996**, *37*, 145.
40. Wink, D.A.; Cook, J.A.; Kim, S.Y.; Vodovotz, Y.; Pacelli, R.; Krishna, M.C.; Russo, A.; Mitchell, J.B.; Jourd'heuil, D.; Miles, A.M.; Grisham, M.B. *J. Biol. Chem.*, **1997**, *272*, 11147-11151.
41. Mordan, L.J.; Bertram, J.S. *Cancer Res.,* **1983**, *43*, 567-571.

Chapter 6

Carotenoids, as Promising Factor for Functional Foods

H. Nishino

Department of Biochemistry, Kyoto Prefectural University of Medicine, Kawaramachi-Hirokoji, Kamigyoku, Kyoto 602, Japan

Various natural carotenoids, besides β-carotene, were proven to have anticarcinogenic activity, and some of them, such as α-carotene, showed higher potency than β-carotene. Furthermore, some of these natural carotenoids show more potent antioxidative activity than β-carotene. Thus, these carotenoids (α-carotene, lutein, zeaxanthin, lycopene, β-cryptoxanthin and phytoene), as well as β-carotene, may be applicable as active components in functional foods for the purpose of preventing oxidative damage-related diseases, including cancer. In the case of phytoene, one of the anticarcinogenic carotenoids, the concept of "bio-chemoprevention", which means biotechnology-assisted method for chemoprevention, will be fitted. In Fact, the establishment of mammalian cells producing phytoene was accomplished by introducing the *crt B* gene, which encodes phytoene synthase. These cells were proven to acquire the resistance against the oxidative stress. Phytoene-containing animal foods may be classified as a novel type of functional food, which has the preventive activity against diseases. It may also have the ability to reduce the accumulation of oxidized substances, which are hazardous for human health.

Information has been accumulated indicating that diets rich in vegetables and fruits can reduce the risk of a number of chronic diseases, including cancer, cardiovascular disease, diabetes and age-related macular degeneration. Various factors in plant foods, such as carotenoids, antioxidative vitamins, phenolic compounds, terpenoids, steroids, indoles and fibers, have been considered responsible for this reduction in risk. Among them, carotenoids have been widely studied and proven to show diverse beneficial effects on human health. Initially, carotenoids in vegetables and fruits were suggested to serve as precursors of vitamin A; Vitamin A being the active compound. In this context, β-Carotene has been studied most extensively, since β-Carotene has the highest pro-vitamin A activity among carotenoids. However, Peto et al.(1) suggested that β-Carotene could have a protective effect against cancer without converting to vitamin A. Logically, then, carotenoids other than β-Carotene may also contribute to protection of cancer and various diseases.

Of more than 600 carotenoids identified up to date, about 40 carotenoids are found in our daily foods. However, as a result of selective uptake in the digestive tract, only 14 carotenoids with some of their metabolites have been identified in human plasma and tissues. Thus, it was decided to evaluate the biological activities of only the carotenoids that are detectable in the human body. It was found that some of the carotenoids that were tested showed more potent activity than β-Carotene to suppress the process of carcinogenesis. In addition, antioxidative activity of α-carotene and lycopene was also proven to be higher than that of β-Carotene. Therefore, various natural carotenoids, besides β-Carotene, seem to be possible candidates as factors for functional foods.

Initially, recent results of the evaluation for anticarcinogenic activity of the various natural carotenoids will be summarized, as the part of the basic data for the development of functional foods. Some of the natural carotenoids, such as phytoene, are unstable when purified, and thus, very difficult to examine the biological activities of them. In this context, we attempted to develop a new method for the synthesis of phytoene in animal cells. Discovery of mammalian cells producing phytoene was succeeded by the introduction of *crt B* gene, which encodes phytoene synthase. These cells were proven to acquire the resistance against oxidative stress. Thus, the antioxidative activity of phytoene was proven.

Anti-Carcinogenic Activity of Natural Carotenoids

Among the carotenoids, α-carotene, lutein, zeaxanthin and lycopene are now being investigated, by international collaboration, as promising candidates for cancer prevention. This widespread investigation is, in part, due to the fact that these carotenoids are detectable in human plasma, and may have a more potent anticarcinogenic activity than β-carotene.

α-Carotene. In recent studies, it was found that α-carotene induced the G1-arrest in the process of the cell cycle (2). Since various agents which induce G1-arrest have been proven to have cancer preventive activity, we evaluated anticarcinogenic activity of α-carotene. α-Carotene showed higher activity than β-carotene to suppress tumorigenesis in skin, lung, liver, and colon (3-4).

In the skin tumorigenesis experiment, a two-stage mouse skin carcinogenesis modes were used. Seven-week-old ICR mice had their backs shaved with an electric clipper. Starting one week after initiation with 100μg of 7,12-dimethylbenz[a]anthracene (DMBA), 1.0μg of 12-*O*-tetradecanoylphorbol-13-acetate (TPA) was applied twice a week for 30 weeks. α- or β-carotene (200 nmol) was applied with each TPA application. With this, greater potency of α-carotene over β-carotene was observed. The percentage of tumor-bearing mice in the control group was 68.8%, whereas the percentages of tumor-bearing mice in the groups treated with α- and β-carotene were 25.0% and 31.3% respectively. The average number of tumors per mouse in the control group was 3.73, whereas the α-carotene-treated group had 0.27 tumors per mouse ($p<0.01$, Students T-test). β-carotene treatment also decreased the average number of tumors per mouse (2.94 tumors per mouse), but the difference from the control group was not significant.

The greater potency of α-carotene than β-carotene in the suppression of tumor promotion was confirmed by another two-stage carcinogenesis experiment. This experiment used a 4-nitroquinoline-1-oxide (4NQO)-initiated and glycerol-promoted ddY mouse lung carcinogenesis model. 4NQO (10mg/kg body weight) was given by a single s.c. injection on the first experimental day. Glycerol (10% in drinking water) was given from experimental week 5 to week 30, continuously. α- and β-carotene (at the concentration of 0.05%), or vehicle as a control was mixed as an emulsion into drinking water during the promotion stage. The average number of tumors per mouse in the control group was 4.06, whereas the α-carotene-treated group had 1.33 tumors per mouse ($p<0.001$). β-Carotene treatment did not show any suppressive effect on the average number of tumors per mouse, but rather induced a slight increase (4.93 tumors per mouse).

In the liver carcinogenesis experiment, a spontaneous liver carcingogenesis model was used. Male C3H/He mice, which have higher incidences of spontaneous liver tumor development, were treated for 40 weeks with α- and β-carotene (at the concentration of 0.05%, mixed as an emulsion into the drinking water) or vehicle as a control. The mean number of hepatomas was significantly decreased by the α-carotene treatment as compared with that in the control group(the control group developed 6.31 tumors per mouse, whereas the α-carotene-treated group had 3.00 tumors per mouse; $p<0.001$). On the other hand, the β-carotene-treated group did not show a significant difference from the control group, although a tendency toward a decrease was observed (4.71 tumors per mouse).

As a short term experiment to evaluate the suppressive effect of α-carotene on colon carcinogenesis, the effect of N-methylnitrosourea (MNU, three intrarectal administrations of 4mg in week 1)-induced colonic aberrant crypt foci formation was examined in Sprague-Dawley (SD) rats. α- or β-Carotene (6 mg, suspended in 0.2 ml of corn oil, intergastric gavage daily), or vehicle as control, were administered during weeks 2 and 5. The mean number of colonic aberrant crypt foci in the control group was 62.7, whereas the α- or β-Carotene-treated group had 42.4 (significantly different from the control group: $p<0.05$) and 56.1 respectively. Thus, the greater potency of α-carotene over β-carotene was, again, observed.

Lutein. Lutein is the dihydroxy form of α-carotene and is distributed among a variety of vegetables such as kale, spinach, and winter squash. It may also be found in fruits such as mango, papaya, peaches, prunes, and oranges.   An epidemiological study in the Pacific Islands indicated that people with a high intake of all three of β-carotene, α-carotene, and lutein had the lowest risk of lung cancer (5). Thus, the effect of lutein on lung carcinogenesis was examined.

Lutein showed anti-tumor promoting activity in a two-stage carcinogenesis experiment in the lungs of ddY mice. This process was initiated with 4NQO and promoted with glycerol. 4NQO (10 mg/kg body weight), dissolved in a mixture of olive oil and cholesterol (20:1), was given by a single s.c. injection on the first experimental day. Glycerol 910% in drinking water) was given as a tumor promoter from experimental week 5 to week 30, continuously. Lutein, 0.2 mg in 0.2 ml of mixture of olive oil and Tween 80 (49:1), was given by oral intubation three times a week during the tumor promotion stage (25 weeks). Treatment with lutein showed a tendency to decrease lung tumor formation. The control group developed 3.07 tumors per mouse, whereas the lutein-treated group had 2.23 tumors per mouse. ***

To confirm the antitumor promoting activity of lutein another two-stage carcinogenesis experiment was conducted. This proved the anti-tumor promoting activity of lutein in a two-stage carcinogenesis experiment in the skin of ICR mice. Initiation of tumor formation was done with DMBA and promoted with TPA and mezerein. At 1 week after initiation by 100 μg of DMBA, TPA (10 nmol) was applied once, then mezerein (3 nmol for 15 weeks, and 6 nmol for a subsequent 15 weeks) twice a week. Lutein (1 μmol, molar ratio to TPA = 100) was applied twice (45 minutes before, and 16 hours after the TPA application). At the experimental week 30, the average number of tumors per mouse in the control group was 5.50, whereas the lutein-treated group had 1.91 tumors per mouse (p<0.05) (Nishino, H., J. Cell Biochem., 1997, in press).

Lutein also inhibited the development of aberrant crypt foci in SD rat colon, induced by MNU (three intrarectal administrations of 4mg in week 1). Lutein (0.24 mg, suspended in 0.2 ml of corn oil, intergastric gavage daily), or vehicle as control, was administered during weeks 2 and 5. The mean number of colonic aberrant crypt foci in the control group at week 5 was 69.3, whereas the lutein-treated group had 40.2 (significantly different from control group: p<0.05) (4).

Zeaxanthin. Zeaxanthin is the dihydroxy form of β-carotene, and is distributed in our daily foods such as corn, and various other vegetables. Since the awareness of zeaxanthin as being a beneficial carotenoid has been quite recent, available data for zeaxanthin is minimal.

Recently, some features of zeaxanthin were elucidated. For example, zeaxanthin suppressed TPA-induced expression of an early antigen of the Epstein-Barr virus in Raji cells (Fig. 1). TPA-enhanced ^{32}Pi-incorporation into phospholipids of cultured cells was also inhibited by zeaxanthin Fig. 2). Anticarcinogenic activity of zeaxanthin *in vivo* was also examined. For example, it was found that spontaneous liver carcinogenesis in C3H/He male mice was suppressed by the treatment with zeaxanthin (at the concentration of 0.005%, mixed as an emulsion into drinking water) as shown in Table I.

Lycopene. Lycopene occurs in our diet predominantly in tomatoes and tomato products. An epidemiological study in elderly Americans has indicated that a high

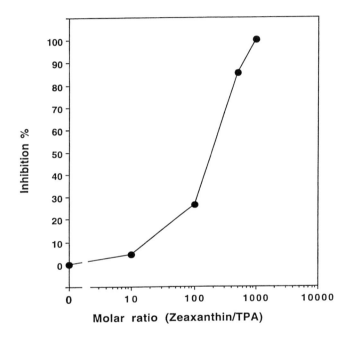

Fig. 1. Effect of zeaxanthin on TPA-induced expression of early antigen of Epstein-Barr virus in Raji cells.

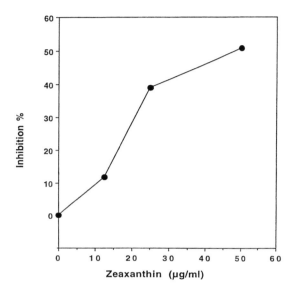

Fig. 2. Effect of zeaxanthin on TPA-enhanced ^{32}Pi-incorporation into phospholipids of cultured HeLa cells.

Table I. Effect of Zeaxanthin on Tumorigenesis in Mouse Liver

Group	Number of Mice	Tumor-bearing Mice (%)	Average Number of Tumors per
Control	14	35.7	1.75
+Zeaxanthin	12	8.3	0.08

Zeaxanthin, 0.005% in drinking water, was administered to male C3H/HC mice for a period of 40 weeks.

tomato intake has been associated with a 50% reduction in mortality from cancers at all sites (6). A case control study conducted in Italy showed a potential protection, with a high consumption of lycopene in the form of tomatoes, against cancers of the digestive tract (7). An inverse relationship between a high intake of tomato products and prostate cancer risk has also been reported, along with a study showing lycopene's exceptionally high singlet oxygen carrying capacity(8-10).

Studies concerning the anticarcinogenic activity of lycopene in animal models were carried out in the mammary gland, liver, lung, skin, and colon (4, 11, Nishino, H., J. Cell Biochem., 1997, in press). This study in mice, with a high rate of spontaneous mammary tumors, showed that intake of lycopene delayed and reduced tumor growth.

Spontaneous liver carcinogenesis in C3H/He male mice was also suppressed by exposure to lycopene. Treatment for 40 weeks with lycopene (at the concentration of 0.005%, mixed as an emulsion into drinking water) resulted in a significant decrease of liver tumor formation; the control group developed 7.65 tumors per mouse, whereas the lycopene-treated group had 0.92 tumors per mouse ($p<0.005$).

Lycopene has also shown antitumor promoting activity in a two-stage carcinogenesis experiment in the lung of ddY mice. This process was initiated with 4NQO, and promoted with glycerol. 4NQO (10 mg/kg body weight), dissolved in a mixture of olive oil and cholesterol (20:1), was given by a single s.c. injection on the first experimental day. Glycerol (10% in drinking water) was given as tumor promoter from experimental week 5 to week 30, continuously. Lycopene, 0.2 mg in 0.2 ml of mixture of olive oil and Tween 80 (49:1), was given by oral intubation three times a week during the tumor promotion stage (25 weeks). Treatment with lycopene resulted in the significant decrease of lung tumor formation. The control group developed 3.07 tumors per mouse, whereas the lycopene-treated group had 1.38 tumors per mouse ($p<0.05$).

The anti-tumor promoting activity of lycopene was confirmed by another two-stage carcinogenesis experiment. It showed anti-tumor promoting activity in the skin of ICR mice. This process was initiated with DMBA and promoted with TPA. From 1 week after initiation by 100 µg of DMBA, 1.0 µg (=1.6 nmol) of TPA was applied twice a week for 20 weeks. Lycopene (160 nmol, molar ratio to TPA = 100) was applied with each TPA application. At the experimental week 20, the average number of tumors per mouse in the control group was 8.53, whereas the lycopene-treated group had 2.13 tumors per mouse 9$p<0.05$).

Lycopene also inhibited the development of aberrant crypt foci in SD rat colon induced by MNU (three intrarectal administration of 4 mg in week 1). Lycopene (0.12 mg, suspended in 0.2 ml of corn oil, intergastric gavage daily), or vehicle as control, were administered during weeks 2 and 5. The mean number of colonic aberrant crypt foci in control group at week 5 was 69.3, whereas the lycopene-treated group had 34.3 (significantly different from control group: $p<0.05$).

Other Carotenoids. In addition to the carotenoids mentioned above, β-cryptoxanthin seems to be another promising carotenoid. It has shown the strongest inhibitory activity in the *in vitro* screening test (12). β-Cryptoxanthin is distributed in our daily foodstuff, such as oranges, and is one of the major carotenoids which is detectable in human blood. Thus, it seems worthy to investigate more precisely.

Production of Phytoene in Mammalian Cells

Establishment of Phytoene Producing Mammalian Cells, and Analysis of their Properties. Phytoene, which is detectable in human blood, was proven to suppress tumorigenesis in the skin. It was suggested that the antioxidative activity of phytoene may play an important role in its action mechanism. In order to confirm the mechanism, a more precise study should be carried out. However, phytoene is quite unstable when it is purified, thus, becomes difficult to examine the biological activities of phytoene. To aid in the evaluation of its biological properties, stable production of these carotenoids in target cells was attempted. The phytoene synthase-encoding gene, *crt B*, has already been cloned from *Erwinia uredovora* (13), thus it was used for the expression of the enzyme in animal cells.

Mammalian expression plasmids, pCAcrtB, to transfer the *crt B* to mammalian cells, were constructed as follows. First, the sequence around the initiation codon of the *crt B* gene on the plasmid pCRT-B, was modified by PCR. This was done by using the primers to replace the original bacterial initiation codon TTG with CTCGAGCCACCATG. This is a composite of the typical mammalian initiation codon ATG preceded by the Kozak consensus sequence and a XhoI recognition site. The XhoI linker which harbors a cohesive end for the EcoR1 site was ligated to the EcoR1 site at the 3'-end of the *crt B* gene. The 969-gase pair (bp) XhoI fragment was cloned into the XhoI site of the expression vector pCAGGS. The resulting plasmid pCAcrtB drives the *crt B* gene by the CAG promoter (modified chicken β-actin promoter coupled with cytomegalovirus immediate early enhancer). In the pCAGGS vector, a rabbit β-globin polyadenylation signal is provided just downstream of the XhoI cloning site.

Plasmids were transfected either by electroporation, or lipofection. For the gene transfer to NIH3T3 cells (cultured in Dulbecco's modified minimum essential medium (DMEM) supplemented with 4 mM L-glutamin, 80 U/ml penicillin, 80 mg/ml streptomycin and 10% calf serum (CS)), the parameter for electroporation using a Gene Pulser (BioRad) was set at 1,500 V/25mF with a DNA concentration of 12.5-62.5 μg/ml. Lipofection was carried out using LIPOFECTAMINE (GIBCO BRL) according to the protocol supplied by the manufacturer. pCAcrtB or pCAGGS was cotransfected with the plasmid pKOneo, which harbors a neomycin resistance-encoding gene (kindly provided by Dr. Douglas Hanahan, University of California, San Francisco).

For Northern blot analysis, 20 μg of total RNA was loaded onto a 1.2% formaldehyde agarose gel, electrophoresed and transferred to a nitrocellulose filter (Nitroplus). The 969-bp XhoI fragment of the *crt B* gene as mentioned above was labeled with [^{32}P]dCTP by the random primer labeling method and used as a probe to hybridize the target RNA on the filter. NIH#T# cells transfected with pCAcrtB showed the expression of a 1.5 kilobases mRNA from the *crt B* gene as a major transcript. Those transcripts were not present in the cells transfected with the vector alone.

For analysis of phytoene by HPLC, the lipid fraction, including phytoene, was extracted from cells (10^7 - 10^8). The sample was subjected to HPLC (column: 3.9 by 300mm, Nova-pakHR, 6m C18, Waters) at a flow rate of 1 ml/min. To detect phytoene, UV absorbance of the eluate ate 286 nm was measured by a UV detector (JASCO875). Phytoene was detected as a major peak in HPLC profile in NIH3T3 cells transfected with pCcrtB, but not in control cells. Phytoene was identified by UV and field desorption mass-spectra.

Since lipid peroxidation is considered to play a critical role in tumorigenesis, and it was suggested that the antioxidative activity of phytoene may play an important role in its mechanism of anticarcinogenic action, the level of phospholipid peroxidation induced by oxidative stress in cells transfected with pCAcrtB or with vector alone was compared.

Oxidative stress was imposed by culturing the cells in a Fe^{3+}/adenosine 5'-diphosphate (ADP) containing medium (374 mM iron (III) chloride, 10mM ADP dissolved in DMEM) for 4 hr. The cells were then washed three times with a Ca^{2+} and Mg^{2+}-free phosphate buffered saline (PBS(-)) harvested by scraping, washed once with PBS(-), suspended in 1 ml of PBS(-), and freeze-thawed once. The lipid fraction was then extracted from the cell suspension twice with 6 ml of chloroform/methanol (2:1). The chloroform layer was collected, and dried with sodium sulfate. The sample was evaporated, its residue was dissolved in a small volume of HPLC solvent (2-propanol:n-hexane:methanol:water = 7:5:1:1), and then subjected to chemiluminescence HPLC (CL-HPLC).

The lipids were separated with the column (Finepack SIL NH2-5,250mm X 4.6mm i.d. JASCO) by eluting with a HPLC solvent (see above) at a flow rate of 1 ml/min at 35°C. Post column chemiluminescent reaction was carried out in the mixture of ctyochrome c (10 mg/ml) and luminol (2 mg/ml) in a borate buffer (pH 10) at a flow rate of 1.1 ml/min. To detect lipids, UV absorbance of the eluate at 210 nm was measured by a UV-8011 detector (TOSOH). Chemiluminescence was detected with a CLD-110 detector (Tohoku Electric Ind.). The phospholipid hydroperoxidation level in the cells that had been transfected with pCAcrtB and confirmed to produce phytoene by HPLC, was lower than that in the cells transfected with vector alone (Fig. 3). Thus, anti-oxidative activity of phytoene in animal cells was confirmed.

It is of interest to test the effects of the endogenous synthesis of phytoene on the malignant transformation process, which is newly triggered in non-cancerous cells. Thus, a study was carried out on the NIH3T3 cells producing phytoene for its possible resistance against oncogenic insult imposed by transfection of the activated H-ras oncogene. Plasmids with activated H-ras gene was transfected to NIH3T3 cells with or without phytoene production, and the rate of transformation focus formation in 100mm diameter dishes was compared. It was proven that the rate of transformation focus formation induced by the transfection of activated H-ras oncogene was lower in the phytoene producing cells than in control cells.

This type of experimental method may be applicable for the evaluation of the anti-carcinogenic activities of other phytochemicals, since the cloning of genes for

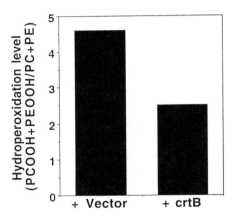

Fig. 3. Reduction of oxidative stress-induced lipid hydroperoxidation level in NIH3T3 cells producing phytoene.

the synthesis of various kinds of substances in vegetables and fruits has already been accomplished. It is particularly useful to evaluate the biological activity of unstable phytochemicals, such as phytoene and other carotenoids.

Bio-chemoprevention. Valuable chemopreventive substances, including carotenoids, may be produced in a wide variety of foods by means of biotechnology. This new concept may be named as "bio-chemoprevention". In this prototype experiment, phytoene synthesis in animal cells is demonstrated. Since phytoene produced in animal cells was proven to prevent oxidative damage of cellular lipids, it may become a valuable factor in animal foods to reduce the formation of oxidized oils, which are hazardous for health. It may also aid in keeping freshness, resulting in the maintenance of a good quality of foods. Furthermore, phytoene-containing foods are valuable in the prevention of cancer, since phytoene is known as an anticarcinogenic substance. Thus, phytoene synthesis may become one of the fundamental methods for the development of novel animal foods.

Acknowledgments. This work was supported in part by grants from the Program for Promotion of Basic Research Activities for Innovative Biosciences, the Program of Fundamental Studies in Health Sciences of the Organization for Drug ADR Relief, R&D Promotion and Product Review, the Ministry of Health and Welfare (The 2nd-term Comprehensive 10-year Strategy for Cancer Control), the Ministry of Education, Science and Culture, SRF, and the Plant Science Research Foundation, Japan. The study was carried out in collaboration with research groups of Kyoto Prefectural University of Medicine, Akita University College of Allied Medical Science, Kyoto Pharmaceutical University, National Cancer Center Research Institute, Dainippon Ink & Chemicals, Inc., Lion Co. and Kirin Brewery Co., Japan, and LycoRed Natural Products Industries, Ltd., Israel, and Dr. Frederick Khachik, U.S. Department of Agriculture, U.S.A.

Literature Cited

1. Peto, R.; Doll, R.; Buckley, J.D.; Sporn, M.B. *Nature* 1981, 290, 201-208.
2. Murakoshi, M.; Takayasu, J.; Kimura, O.; Kohmura, E.; Nishino, H.; Iwashima, A.; Okuzumi, J.; Sakai, T.; Sugimoto, T.; Imanishi, J.; Iwasaki, R. *J. Natl. Cancer Inst.* 1989, 81, 1649-1652.
3. Murakoshi, M.; Nishino, H.; Satomi, Y.; Takayasu, J.; Hasegawa, T; Tokuda, H.; Iwashima, A.; Okuzumi, J.; Okabe, H.; Kitano, H.; Iwasaki, R. *Cancer Res.* 1992, 52, 6583-6587.
4. Narisawa, T.; Fukaura, Y.; Hasebe, M.; Ito, M.; Aizawa, R.; Murakoshi, M.; Uemura, S.; Khachik, F.; Nishino, H. *Cancer Lett.* 1996, 107, 137-142.
5. Le Marchand, L.; Hankin, J.H.; Kolonel, L.N.; Beecher, G.R.; Wilkens L.R.; Zhao, L.P. *Cancer Epidemiol. Biomarkers Prev.* 1993, 2, 183-187.
6. Colditz, G.A.; Branch, L.G.; Lipnick, R.J. *Am. J. clin. Nutr.* 1985, 41, 32-36.
7. Franceschi, S.; Bidoli, E.; La Veccia, C.; Talamini, R.; D'Avanzo, B.; Negri, E. *Int. J. Cancer* 1994, 59, 181-184.

8. Giovannucci, E.; Ascherio, A.; Rimm, E.B.; Stampfer, M.J.; Colditz, G.A.; Willett, W.C. *J. Natl. Cancer Inst.* 1995, 87, 1767-1776.

9. Stahl, W.; Sies, H. *Ann. N.Y. Acad. Sci.* 1993, 691, 10-19.

10. Ukai, N.; Lu, Y.; Etoh, H.; Yagi, A.; Ina, K.; Oshima, S.; Ojima, F.; Sakamoto, H.; Ishiguro, Y. *Biosci. Biotech. Biochem.* 1994, 58, 718-1719.

11. Nagasawa, K.; Mitamura, T.; Sakamoto, S.; Yamamoto, K. *Anticancer Res.* 1995, 15, 1173-1178.

12. Tsushima, M.; Maoka, T.; Katsuyama, M.; Kozuka, M.; Matsuno, T.; Tokuda, H.; Nishino, H.; Iwashima, A. *Biol. Pharm. Bull.* 1995, 18, 227-233.

13. Misawa, N.; Nakagawa, M.; Kobayashi, K.; Yamano, S.; Izawa, Y.; Nakamura, K.; Harashima, K. *J. Bacteriol.* 1990, 172, 6704-6712.

Chapter 7

Metabolism of Dietary Carotenoids and Their Potential Role in Prevention of Cancer and Age-Related Macular Degeneration

F. Khachik[1,2], L. A. Cohen[3], and Z. Zhao[3]

[1]Department of Chemistry, The Catholic University of America, 620 Michigan Avenue, NE, Washington DC 20064
[2]Department of Chemistry and Biochemistry, University of Maryland, College Park, MD 20742
[3]Nutrition and Endocrinology Section, American Health Foundation, Valhalla, NY 10595

Although the number of dietary carotenoids in fruits and vegetables is in excess of 40, only 13 from selected classes are absorbed, metabolized, and utilized by the human body. Vitamin A active carotenoids, such as α- and β-carotene are absorbed intact and in part, converted to Vitamin A. In contrast, metabolic transformations of other major carotenoids such as lutein, zeaxanthin, and lycopene involve a series of oxidation-reduction reactions. Thirty-four carotenoids, including 13 stereoisomers, and 8 metabolites have been characterized in human serum and milk employing two HPLC methods. Similarly, in addition to the presence of lutein and zeaxanthin in human retina, reported previously, their oxidative metabolites have recently been isolated from dissected human retina. An *in vivo* preclinical bioavailability and toxicity study with lycopene involving rats is described. Proposed metabolic pathways of carotenoids and their potential role in prevention of cancer and age-related macular degeneration are discussed.

Although approximately 600 carotenoids have been isolated from natural sources and characterized (1), the number of dietary carotenoids in common fruits and vegetables consumed in the U.S. is in excess of 40 (2,3). Only 13 all-E-carotenoids, which belong to the class of hydroxy- and hydrocarbon carotenoids (carotenes), have been detected in human serum and milk (4). In addition, eight metabolites resulting from two major dietary carotenoids, namely, lutein and lycopene have also been characterized (4-9). The Vitamin A active carotenoids in human serum or plasma and milk are: α-carotene, β-carotene, β-crypotxanthin, and γ-carotene. However, the nutritional significance of other non-Vitamin A active carotenoids in prevention of cancer (10,11), heart disease (12,13), age-related macular degeneration, or AMD (14) has also been realized.

Here we describe an updated summary of the distribution of carotenoids and their metabolites in human serum, milk, and retina. We also describe a recent preclinical bioavailability study with lycopene, involving rats, and propose pathways for the *in vivo* metabolic oxidation of this compound. Based on this study and our previously reported human supplementation studies with purified carotenoids, the antioxidant mechanism of action for lycopene, lutein, and zeaxanthin as potential candidates for lowering the risk for cancers and AMD is discussed.

Distribution of Carotenoids in Human Serum and Milk

Carotenoids in human serum and milk originate from consumption of fruits and vegetables, which are one of the major dietary sources of these compounds. The carotenoids of fruits and vegetables can be classified as: 1) hydrocarbon carotenoids or carotenes, 2) monohydroxycarotenoids, 3) dihydroxycarotenoids, 4) carotenol acyl esters, and 5) carotenoid epoxides. Among these classes, only carotenes (group 1), monohydroxy-(group 2) and dihydroxycarotenoids (group 3) are found in the human serum and milk. Carotenol acyl esters (group 4) apparently undergo hydrolysis in the presence of pancreatic secretions to regenerate their parent hydroxycarotenoids, which are then absorbed. Although carotenoid epoxides (group 5) have not been detected in human serum or tissues, and their fate is uncertain at present, a recent *in vivo* bioavailability study with lycopene, involving rats, indicates that this class of carotenoids may be handled and modifed by the liver. This study will be described, in detail, later in this text. For a comprehensive review of the distribution of carotenoids in fruits and vegetable, see the articles by Khachik et al. (15-17). The concentration of carotenoids in fruits and vegetables commonly consumed in the U.S. has also been reported (17-20). Comparison of the qualitative profile of carotenoids in foods with those of human serum and milk is shown in Figure 1. The first group is Vitamin A active carotenoids, such as α-carotene, β-carotene, β-crypotxanthin, and γ-carotene.

The second group: α-cryptoxanthin, neurosporene, ζ-carotene, phytofluene, and phytoene, have no Vitamin A activity and appear to be absorbed intact. At present, there is no evidence to suggest that these carotenoids undergo metabolic transformation.

Several oxidative metabolites of lycopene, lutein, zeaxanthin, and lactucaxanthin in human serum and milk have been isolated and characterized (4). Human bioavailability and metabolic studies have supported the possibility of *in vivo* oxidation of these carotenoids in humans (9,21). The metabolic transformation of these carotenoids involves a series of oxidation-reduction reactions, which have been described previously (9,21). There are also two metabolites of lutein which are apparently formed in the presence of acids by the non-enzymatic dehydration of this compound in the human digestive system (8).

As mentioned earlier, carotenol acyl esters are abundant in many fruits and vegetables (2), but have not been detected in human serum or milk (4,5). Only two monohydroxy carotenoids, namely, α-cryptoxanthin and β-cryptoxanthin have been

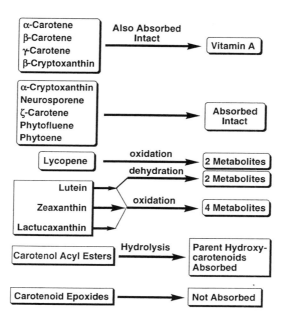

Figure 1. Comparison between dietary and serum carotenoids and their metabolic transformation in humans.

detected in common fruits and vegetables, as well as human serum and milk. Similarly, of all the dihydroxycarotenoids isolated from various natural sources, only lutein, zeaxanthin, and lactucaxanthin have been found in foods, human serum and milk (4,14). The concentration of lactucaxanthin in human plasma is normally very low, since the dietary source of this compound is limited to date. We have detected it only in Romain lettuce, *Lactuca sativa* (15).

A complete list of dietary carotenoids detected in human serum and milk is provided in Table I. These are 25 carotenoids consisting of 13 all-E-, and 12 Z-compounds. In addition, there are 1 Z- and 8 all-E-carotenoid metabolites in human serum and milk. The structures of these metabolites are shown in Figure 2. The only mono-Z-carotenoid metabolite identified in human serum and milk to date is that of 3-hydroxy-β,ε-caroten-3'-one. The location of the Z-bond in this compound is not known at present. Employing two HPLC methods, all of these 34 carotenoids and their metabolites can be readily separated and quantified (4).

Lutein, Zeaxanthin, and Their Oxidative Metabolites in Human Retina

In 1985, for the first time, Bone et al. (22) presented preliminary evidence that the human macular pigment is a combination of lutein and zeaxanthin. More recently, in 1993, Bon et al. (23) elegantly established the complete identification and stereochemistry of the human macular pigment as lutein [(3R,3'R,6'R)-β,ε-carotene-3,3'diol], zeaxanthin [(3R,3'R)-β,β-carotene-3,3'diol] and (3R,3'S,meso)-zeaxanthin [3R,3'S)-β,β-carotene-3,3'diol]. In a case-control study published in 1994, high consumption of fruits and vegetables, rich specifically in lutein and zeaxanthin, has been correlated with a lower risk for age-related macular degeneration (AMD) (14). The results from that study indicated that persons consuming green leafy vegetables containing a high concentration of lutein (ca.6 mg/day) had a 43% lower risk of exudative AMD, compared to the subjects in the lowest quintile.

Recently, we have separated and characterized three major and 11 minor carotenoids in human retina by HPLC-UV-Vis-MS (24). The major carotenoids have been identified as lutein, zeaxanthin, and a direct oxidation product of lutein, namely, 3-hydroxy-β-ε-caroten-3'-one (Figure 2). Several oxidation products of lutein, zeaxanthin, and one of lycopene (2,6-cyclolycopene-1,5-diol I) whose structures are shown in Figure 2, were also among the minor carotenoids. In addition, the most common geometrical isomers of lutein and zeaxanthin, i.e. 9Z-lutein, 9'Z-lutein, 13Z-lutein, 13'Z-lutein, 9Z-zeaxanthin, and 13Z-zeaxanthin, normally present in serum, were also detected at low concentrations in retina. Similar results were also obtained from an extract of freshly dissected monkey retina. At present, this finding provides the only data, possibly suggesting *in vivo* metabolic oxidation of lutein in human retina. The protection of retina from short-wavelength visible light by lutein and zeaxanthin is based on two assumptions; 1) the oxidation products of lutein and zeaxanthin are formed *in vivo* in retina and 2) these oxidative metabolites are formed by the action of blue light. Based on supplementation studies with purified lutein and zeaxanthin involving human subjects, we previously proposed metabolic pathways for conversion of these dietary carotenoids to their oxidation products (9,21). According to these metabolic transformations,

Table I. Dietary Carotenoids in Human Serum and Milk

Entry	Carotenoids	Chemical Class
	all-E-Dietary Carotenoids[a]	
1	α-carotene[a]	Hydrocarbons (carotenes)
2	β-carotene[a]	
3	γ-carotene[a]	
4	lycopene	
5	neurosporene	
6	ζ-carotene	
7	phytofluene	
8	phytoene	
9	α-cryptoxanthin	Monohydroxycarotenoids
10	β-cryptoxanthin[a]	
11	lutein	Dihydroxycarotenoids
12	zeaxanthin	
13	lactucaxanthin	
	Z-Dietary Carotenoids[a]	
14	9Z-β-carotene[a]	Hydrocarbons (carotenes)
15	13Z-β-carotene[a]	
16	Z-lycopenes[b]	
17	Z-phytofluenes[b]	
18	Z-β-cryptoxanthin[a,b]	Monhydroxycarotenoid
19	13,13'-Z-lutein	Dihydroxycarotenoids
20	9Z-lutein	
21	9'Z-lutein	
22	13Z-lutein+13'Z-lutein	
23	9Z-zeaxanthin	
24	13Z-zeaxanthin	
25	15Z-zeaxanthin	

[a]Identifies carotenoids with vitamin A activity. [b]The location of the Z-bond in Z-lycopenes, Z-phytofluenes, and Z-β-cryptoxanthin is not known at present.

76

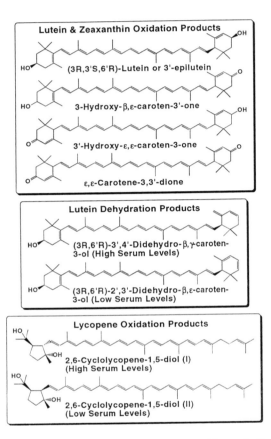

Figure 2. Chemical structure of *all-E*-dietary carotenoid metabolites identified in human serum and milk. Only the absolute configuration rather than the planar structure for those carotenoids with confirmed chirality have been depicted. Only the relative not absolute configuration at C-2, C-5, and C-6 for 2,6-cyclolycopene-1,5-diol I and II is known. With permission from reference 9.

(3R,3'S,meso)-zeaxanthin, (3R,3'R,6'R)-lutein, and (3R,3'R)-zeaxanthin may be inter-converted through a series of oxidation-reduction and double bond isomerization reactions as shown in Fig. 3. We have recently determined that 98% of zeaxanthin in human plasma is identical to the form in the diet, (3R,3'R)-zeaxanthin; only 2% exists as (3R,3'S,meso)-zeaxanthin (Khachik, unpublished results). Since Bone et al. (23) have demonstrated that nearly equal amounts of (3R,3'R)-zeaxanthin and (3R,3'S,meso)-zeaxanthin are present in the human macula, the data indicate that the latter compound may be a result of double bond isomerization of dietary (3R,3'R,6'R)-lutein as shown in Fig.3. However, the only data in support of these metabolic transformations is the presence of 3-hydroxy-β,ε-caroten-3'-one,3'-epilutein, and (3R,3'S,meso)-zeaxanthin in retina. The transport, the metabolic interconversions between (3R,3'R,6'R)-lutein, and the two forms of visible light and/or catalyzed by certain proteins. Human macular carotenoids bind to retinal tubulin (25), and other more specific human macular carotenoid binding proteins, are being identified and characterized (Bernstein et al, unpublished results).

In Vivo Studies with Carotenoids in Rodents

Recently, the inhibitory effect of four dietary carotenoids, α-carotene, β-carotene, lycopene, and lutein, prevalent in human blood and tissues against the formation of colonic aberrant crypt foci, has been reported in Sprague-Dawley rats (26). The rats received three intarectal doses of N-mehtylnitrosourea in week 1, and a daily gavage of de-escalated doses of carotenoids during weeks 2 and 5. Lycopene, lutein, α-carotene, and palm carotenes (a mixture of α-carotene, β-carotene, and lycopene) inhibited the development of aberrant crypt foci quantitated at week 6. In contrast, β-carotene was not effective. This study suggests that lycopene and lutein in small doses may potentially lower the risk for colon carcinogenesis.

In a recent epidemiological study, the intake of carotenoids and retinol in relation to the risk of prostate cancer has been examined (27). This study concluded that consumption of foods rich in β-carotene, α-carotene, lutein, and β-cryptoxanthin were not associated with the risk of prostate cancer, while high consumption of foods rich in tomato products had a significant effect in lowering the risk. Since lycopene is the most abundant carotenoid in tomato products (20), the reduction in the risk of prostate cancer was, therefore, related to the cancer preventive properties of this compound. In recent years, we have identified two oxidation products of lycopene in humans and have proposed pathways leading to the formation of these oxidative metabolites (9,21). Furthermore, the metabolites of lycopene, 2,6-cyclolycopene-1,5-diols I and II (Fig. 2), have been suggested to be potentially useful in cancer chemoprevention due to their ability to enhance the expression of a group of gap junctional communication proteins in an *in vitro* model (28).

Bioavailability Study with Lycopene in Rodents

We recently conducted a preclinical toxicity and bioavailability study with lycopene, involving rodents. The purpose of this study was to determine the biological acceptable dose range, maximal tolerated dose (MTD), and tissue disposition of

78

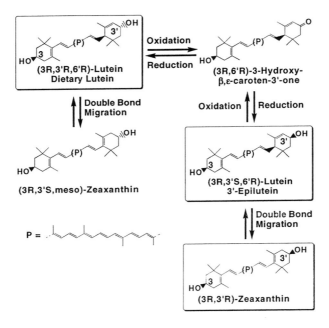

Figure 3. Proposed metabolic transformation of dietary (3R,3'R,6'R)-lutein and (3R,3'R)-zeaxanthin in human retina. With permission from reference 24.

dietary lycopene administered as tomato oleoresin, in AIN-76A diet in male and female rats.

Method and Materials. This was a 10-week pilot study in which the experimental groups of inbred Fischer F344 male and female rats, at about 35 days of age, were randomized into six groups (Table II) to ensure animals in each group were of the same mean weight. A 4% lycopene suspension in medium chain triglycerides, or MCT ("Betatene"), was obtained from Henkel Corporation (LaGrange, IL) and was incorporated into the semipurified AIN-76A diet. All animals, including controls, were given MCT and received the same amount of fat to control for the differences in the lipid content of the various treatment groups. The estimated concentration of lycopene in the diet of each treatment group is shown in Table II. In addition to lycopene, the "Betadene" suspension was shown by High Performance Liquid Chromatography (HPLC) to consist of minor quantities of several carotenoids, as shown in Table III. Our previously published HPLC method on a C18-reversed phase column, was used for the analysis of the diets and all the specimens described here (4).

Study Time Line. All animals at 50 days of age were supplemented with the AIN-76A diet containing the "Betadene" suspension for 70 days; the placebo groups only received the diet and MCT, without lycopene, and carotenoids for the same duration. Throughout the study, the weight gain of the animals were monitored regularly. After ten week, when the rats were at the age of 120 days, blood samples were collected by heart puncture.

Extraction of Diets, Serum, and Tissues. Diets containing lycopene were homogenized in the HPLC solvents with echinenone as an internal standard. After centrifugation, the supernatant was decanted, and the extraction was repeated until the extract was colorless. The combined extracts were then reduced to an appropriate volume for HPLC analysis. Carotenoids were identified from the comparison of their HPLC-UV/Vis-MS profiles with those of authentic standards. Rat sera (ca. 1 ml.) was extracted (see reference 4) using echinenone as an internal standard. Since liver, prostate, and mammary tissues contain lipids, which interfere with the HPLC analysis, these specimens were saponified prior to extraction. Details of extraction and saponification will be described in a later publication.

Results and Discussion. The concentration of lycopene in serum, feces, and all the major tissues, such as lung, liver, mammary and prostate glands was determined by HPLC. Since the complete details of this study will be published elsewhere, here we only present a brief summary of our results.

Serum Uptake of Lycopene. Serum lycopene concentrations for female and male rats are shown in Table IV. These concentrations appear to vary in a non-linear manner, with regard to dose, in both males and females. In females, the highest lycopene concentrations are found in the serum of groups 3 and 4, rather than group 1, as expected. Similarly, among males, the highest serum concentration of lycopene

Table II. Experimental Groups of Rats and the Lycopene Concentration of the Diets in Maximal Tolerated Dose (MTD) Study

Group	Number of Animals Males	Females	Lycopene Concentration in Diet Estimated (Actual)[a] mM/Kg	mg/Kg
1	10	10	2.5 (2.4)	1340 (1280)
2	10	10	1.0 (0.96)	536 (512)
3	10	10	0.5 (0.48)	268 (256)
4	10	10	0.25 (0.24)	134 (128)
5	10	10	0.1 (0.096)	53 (51)
6	20	20	0.0 (0)	0 (0)
Total No.	70	70		

[a]Betatene suspension containing lycopene was added to the AIN-76A diet, on the assumption that consisted of 4% pure lycopene. However based on HPLC analysis, the Betatene suspension contained 3.7% pure lycopene and 2% of other carotenoids such as β-carotene, ζ-carotene, phytofluene, and phytoene.

Table III. Carotenoid Concentration of Rat Diets Supplemented with the Betatene Suspension of Tomato Oleoresin

Carotenoids	Diets, $\mu g/g$ 1	2	3	4	5	6
lycopene	472	348	168	92	22	0
β-carotene	284	138	69	37	8	0
ζ-carotene	1	9	4	2	0	0
phytofluene	58	28	14	8	2	0
phytoene	70	23	0.3	5	1	0
2,6-cyclolycopene-1,5-diol	2	1.4	0.8	0.5	0.1	0

is found in group 3. These results suggest that a homeostatic mechanism, involving hepatic storage metabolism, and release, in a manner similar to Vitamin A regulate serum lycopene levels.

Uptake of Lycopene by Mammary and Prostate Glands. The uptake of lycopene by the mammary (female) and the prostate (male) glands of rats is shown in Table V. Lycopene concentrations in the mammary fat pad of female rats ranged from a low of 140, to a high of 460 ng/g wet weight. In contrast to serum, the uptake of lycopene by the mammary tissue was dose-dependent with respect to dietary intake of lycopene.

Lycopene concentrations in the prostate gland of male rats ranged from a low of 32, to a high of 147 ng/g of wet weight (Table V). A dose-related pattern of lycopene uptake by the prostate of the male rats was apparent. However, on average, the concentration of lycopene in prostate tissue was an order of magnitude lower than mammary tissue.

Uptake of Lycopene and Carotenoids by Liver. Hepatic lycopene concentrations of female and male rats are shown in Table VI. Lycopene was concentrated to a 100-1000 fold greater extent in the liver, compared to serum or other tissues. Lycopene concentrations in the liver of female rats (high of 120 µg/g wet weight). A non-linear dose response curve for lycopene uptake by the liver was exhibited in males, but not in females. In order to obtain a completed carotenoid profile, the livers from 3 animals in each group were pooled for HPLC-UV/Vis-MS analysis. Therefore, the variability between the hepatic concentration of lycopene in female and male rats may be and artifact of small sampling. Particularly interesting, was the absorption of the other carotenoids present in the "Betadene" suspension by the liver of both female and male rats. These were: ζ-carotene, phytofluene, phytoene, all-E-β-carotene, 9Z-β-carotene, and 13Z-β-carotene. Of interest, is the fact that the ratio of concentration of liver phytofluene and phytoene, to lycopene, is considerably higher than that of the "Betadene" suspension, suggesting a selective uptake of these carotenoids by the liver.

Particularly noticeable was the presence of the two oxidative metabolites of lycopene, namely, 2,6-cyclolycopene-1,5-diols I and II, and their precursors 2,6-cyclolycopene-1,5-epoxide I and II in the liver of female and male rats. Although these diols have been previously identified in human serum, their precursor epoxides have not been detected (4,5,9). Unfortunately, these diols and epoxides of lycopene are also present at low concentration in the "Betadene" suspension, and as a result, their presence in the liver may be of dietary origin. However, it is quite likely that lycopene may also undergo an *in vivo* oxidation in the rat liver to form these metabolic by-products. To date, we have been unable to address the metabolism of carotenoid abundance in many fruits and vegetables, have not been detected in human serum. Furthermore, since human livers and tissues cannot be examined in a systematic manner, the results described here indicate that the rats may provide a convenient model in which the metabolism and *in vivo* oxidation of lycopene, as well as other carotenoids, could be studied.

Table IV. Lycopene Concentration of Rat Serum (ng/mL)

Group No.	X ± SD[a]	Median	Range
Female			
1	187 ± 43	205	123-232
2	169 ± 42	180	109-211
3	245 ± 83	210	174-366
4	313 ± 47	308	262-369
5	145 ± 53	152	81-207
6	0	0	0
Male			
1	168 ± 36	160	134-230
2	227 ± 68	225	148-326
3	278 ± 66	285	174-372
4	231 ± 64	215	153-328
5	171 ± 49	177	100-238
6	0	0	0

[a] N= 6

Table V. Lycopene Concentration of Rat Mammary and Prostate Glands (ng/g wet weight)

Group No.	X ± SD[a]	Median	Range
Mammary			
1	309 ± 131	235	232-460
2	200 ± 030	197	172-231
3	215 ± 062	220	139-282
4 (N=4)	229 ± 054	217	181-288
5	174 ± 057	143	139-239
6	0	0	0
Prostate			
1	97 ± 17	99	79-112
2	95 ± 48	83	54-147
3	50 ± 37	35	23-093
4	52 ± 26	52	26-077
5	47 ± 16	47	32-063
6	0	0	0

[a] N= 3

Table VI. Carotenoid Concentration of Livers of Rats Fed Diets Supplemented with Betatene Suspension

Rats	mM/Kg Diet[a]	lycopene	β-carotene	ζ-carotene	phytofluene	phytoene	2,6-cyclolycopene-1,5-diol
					μg/g wet weight		
Female							
1	2.5	120	11	17	106	66	9
2	1.0	64	6	8	48	35	5
3	0.5	66	8	7	50	40	7
4	0.25	49	7	2	46	46	4
5	0.10	42	4	4	33	38	6
6	0.00	0	0	0	0	0	0
Male							
1	2.5	33	3	5	30	23	4
2	1.0	5	1	1	5	4	0.4
3	0.5	60	5	5	34	21	4
4	0.25	3	1	1	--b	--b	0.3
5	0.10	12	1	1	10	10	0.3
6	0.00	0	0	0	0	0	0

[a]Estimated lycopene concentration in the diet; for actual concentration of lycopene and other carotenoids in the diet of each group of rats see Tables II and III, respectively. b Not detected by HPLC; the detection limit of the HPLC photodiode array detector is approximately 0.1 μg/gm.

Conclusion

The majority of epidemiological studies to date have associated the high consumption of carotenoid-rich fruits and vegetables with a lower risk for several human cancers. Unfortunately, the scientific community at large has prematurely assigned this protective effect to a single carotenoid, β-carotene, because of its Vitamin A activity. During the past 14 years, we have reported that fruits and vegetables contain approximately 40-50 carotenoids, some of which are found at markedly greater concentrations than β-carotene. Among these, 34 carotenoids, including 13 stereoisomers and eight metabolites have been identified in human serum and milk. In addition, as described earlier, the *in vivo* anti-tumor activity of several carotenoids against colon carcinogenesis in a rodent model has been reported. It is quite likely that each of these carotenoids, individually, or in concert with one another, may exert their protective role in prevention of human cancers. Selective uptake of carotenoids by specific tissues may also be related to their functional role in disease prevention. This is certainly the case with respect to the accumulation of lutein and zeaxanthin in human retina, and their photo-protective role in prevention of AMD.

Here, we presented a preclinical bioavailability study with lycopene and have shown that rodents can be successfully used to determine the concentrations of this carotene in serum and tissues. In view of the high accumulation of lycopene in the liver of the rats, in comparison to serum and other tissues, it appears that rodents provide us with a reasonably good model to study the metabolism and cancer preventive properties of lycopene, as well as other carotenoids which have been largely neglected to date. As indicated earlier, the metabolites of lycopene, lutein, and zeaxanthin enhance the expression of a group of gap junctional communication proteins in an *in vitro* model of carcinogenesis (28), emphasizing the importance of carotenoid metabolites. Unfortunately, much of the definitive details concerning the biological activities, role, and function of carotenoids in humans remain unexplored due to the focus on β-carotene. Future human and animal studies should concentrate on elucidating the bioavailability, metabolism, function, interaction, and efficacy of the spectrum of dietary carotenoids, as well as their metabolites.

Literature Cited

1. Pfander, H. *In Key to Carotenoids*; Gerspacher, M.; Rychener, M.; Schwabe, R.; Eds.; Birkhauser; Basel, **1987**, pp 11-218.
2. Khachik, F.; Beecher, G.R.; Goli, M.B.; Lusby, W.R. *Pure Appl. Chem.*, **1991**, 71-80.
3. Khachik, F.; Beecher, G.R.; Goli, M.B.; Lusby, W.R. In *Separation and Quantification of Carotenoids in Foods;* Packer, L.; Ed.; Methods of Enzymology; Academic Press; New York, **1992**, Vol. 213A, pp 347-359.
4. Khachik, F.; Spangler, C.J.; Smith, Jr., J.C.; Canfield, L.M.; Steck, A.; Pfander, H. *Anal. Chem.,* **1997**, 69, 1873-1881.
5. Khachik, F.; Beecher, G.R., Goli, M.B.; Lusby, W.R.; Smith, Jr., J.C.; *Anal. Chem.*, **1992**,64, 2111-2122.
6. Khachik, F., Beecher, G.R.; Goli, M.B.; Lusby, W.R.; Daitch, C.E. In Separation and Quantification of Carotenoids in Human Plasma; Packer, L., Ed.; *Methods in Enzymology*; Academic Press: New York, **1992**, Vol 213A, pp 205-219.

7. Khachik, F.; Englert, G.; Daitch, C.E.; Beecher, G.R.; Lusby, W.R.; Tonucci, L.H.; *J. Chromatog. Biomed. Appl.,* **1992**, 582, 153-166.
8. Khachik, F.; Englert, G.; Beecher, G.R.; *J. Chromatogr. Biomed. Appl.,* **1995**, 670, 219-233.
9. Khachik, F.; Steck, A.; Pfander, H. In Bioavailability, Metabolism, and Possible Mechanism of Chemoprevention by Lutein and Lycopene in Humans; Ohigashi, H.; Osawa, T.; Terao, J.; Watanabe, S.; Yoshikawa, T., Eds.; Food Factors for Cancer Prevention; Springer-Verlag: Tokyo, **1997**, In Press.
10. Micozzi, M.S., Ed., In Nutrition and Cancer Prevention; Marcel Dekker: New York, **1989**, pp 213-241.
11. Van Poppel, G. *Eur. J. Cancer,* **1993**, 29A, 1335-1344.
12. Morris, D.L.; Kritchevsky, S.B.; Davis, C.E.; *J. Am. Med. Assoc.,* **1994**, 272, 1439-1441.
13. Gaziano, J.M.; Manson, J.E.; Branch, L.G.; Colditz, G.A.; Willet, W.C.; Buring, J. E., *Ann. Epidemiol.,* **1995**, 5, 255-260.
14. Seddon, J.M.; Ajani, U.A.; Sperduto, R.D.; Hiller, R.; Blair, N.; Burton, T.C.; Farber, M.D.; Gragoudas, E.S.; Haller, j.; Miller, D.T.; Yannuzzi, L.A.; Willet, W., *J. Am. Med. Assoc.,* **1994**, 272, 1413-1420.
15. Khachik, F.; Nir, Z.; Ausich, R.L. In Distribution of Carotenoids in Fruits and Vegetables as a Criterion for the Selection of Appropriate Chemopreventive Agent; Ohigashi, H.; Osawa, T.; Terao, J.; Watanabe, S.; Yoshikawa, T., Eds.; Food Factors for Cancer Prevention; Springer-Verlag: Tokyo, **1997**, In Press.
16. Khachik, F.; Beecher, G.R.; Whittaker, N.F., *J. Agric. Food Chem.,* **1986**, 34, 603-616.
17. Khachik, F.; Beecher, G.R., *J. of Agric. Food Chem.,* **1988**, 36, 929-937.
18. Khachik, F.; Beecher, G.R.; Lusby, W.R., *J Agric. Food Chem.,* **1989**, 37, 1465-1473.
19. Khachik, F.; Goli, M.B.; Beecher, G.R.; Holden, J.; Lusby, W.R.; Tenorio, M.D.; and Barrera, M.R., *J Agric. Food Chem.,* **1992**, 40, 390-398.
20. Tonucci, L.H.; Holden, J.M.; Beecher, G.R.; Khachik, F.; Davis, C.S.; Mulokozi, G.; *J. Agric. Food Chem.,* **1995**, 43, 579-586.
21. Khachik, F.; Beecher, G.R, Smith, J.C., Jr.; *J. Cellular Biochem.,* **1995**, 22, 236-246.
22. Bone, R.A.; Landrum, J.T.; Tarsis, S.L.; *Vision Res.,* **1985**, 25, 1531-1535.
23. Bone, R.A.; Landrum, J.T.; Hime, G.W.; Cains, A.; *Invest. Ophthalmol. Vis. Sci.,* **1993**, 34, 2033-2040.
24. Khachik, F.; Bernstein, P.; Garland, D.L.; *Invest. Ophthalmol. Vis. Sci.,* **1997**, In Press.
25. Bernstein, P.S., Balashov, N.A.; Tsong, E.D.; Rando, R.R.; *Inves. Ophthalmol. Vis. Sci.,* **1997**, In Press.
26. Narisawa, T.; Fukaura, Y.; Hasebe, M.; Ito, M.; Aizawa, R.; Murakoshi, M.; Uemura, S.; Khachik, F.; Nishino, H.; *Cancer Letters,* **1996**, 107, 137-142.
27. Giovannucci, E.; Ascherio, A.; Rimm, E.B.; Stampfer, M.J.; Colditz, G.A.; Willet, W.C.; *J. Natl. Cancer Inst.,* **1995**, 87, 1767-1776.
28. King, T.J.; Khachik, F.; Borkiewicz, H.; Fukushima, L.H.; Morioka, S.; Bertram, J.S.; *Pure Appl. Chekm.,* **1997**, In Press.

Chapter 8

Auraptene, an Alkyloxylated Coumarin from *Citrus natsudaidai* HAYATA, Inhibits Mouse Skin Tumor Promotion and Rat Colonic Aberrant Crypt Foci Formation

Akira Murakami[1], Wataru Kuki[2], Yasuo Takahashi[2], Hiroshi Yonei[2],
Takuji Tanaka[3], Hiroki Makita[3], Keiji Wada[4], Naomi Ueda[4], Masanobu Haga[4],
Yoshimasa Nakamura[5], Yoshimi Ohto[5], Oe Kyung Kim[5], Hajime Ohigashi[5],
and Koichi Koshimizu[1,5]

[1]Department of Biotechnological Science, Faculty of Biology-Oriented Science and
Technology, Kinki University, Iwade-Uchita, Wakayama 649-64, Japan
[2]Research and Development Division, Wakayama Agricultural Processing Research
Corporation, 398, Tsukatsuki Momoyama-Cho, Nagagun, Wakayama 649-61, Japan
[3]First Department of Pathology, Gifu University School of Medicine, 40 Tsukasa-Machi,
Gifu 500, Japan
[4]Faculty of Pharmaceutical Sciences, Health Sciences University of Hokkaido, Ishikari
Tobetsu, Hokkaido 061-02, Japan
[5]Division of Applied Life Sciences, Graduate School of Agriculture, Kyoto University,
Kyoto 606-01, Japan

Auraptene, a coumarin-related compound, has been isolated from the cold-pressed oil of natsumikan (Citrus natsudaidai HAYATA), as an inhibitor of 12-O-tetradecanoylphorbol-13-acetate (TPA)-induced Epstein-Barr virus (EBV) activation in Raji cells. In a two-stage carcinogenesis experiment with 1.6 nmol of TPA and 0.19 μmol of 7,12-dimethyl benz[a]anthracene (DMBA) in ICR mouse skin, topical application of 160 nmol of auraptene significantly reduced tumor incidence and the number of tumors per mouse by 27% and 23%, respectively. Auraptene at 50 μM almost completely suppressed TPA-induced superoxide (O_2^-)and hydroperoxide (ROOH) generation in differentiated HL-60 cells as well as lypopolysaccharide (LPS)/interferon-γ (IFN-γ)-induced nitric oxide (NO) generation in RAW 264.7 cells. Dietary feeding of auraptene at a dose of 100 or 500 ppm inhibited azoxymethane (AOM)-induced rat colonic aberrant crypt foci (ACF) formation in a dose-dependent manner. Oral administration of auraptene (50 - 200 mg/kg body wt.) clearly enhanced the glutathione S-transferase (GST) activity in mouse liver, suggesting that auraptene is an effective chemopreventer in both tumor initiation and promotion phases.

Cancer is the leading cause of death in many developed countries. At present, cancer prevention is accepted as a promising avenue for cancer control. Chemical carcinogenesis is based on the rationale of the successive three stages, i.e., initiation, promotion and progression, in experimental rodents and presumably in humans (*1*). The first step, initiation, is considered to occur in one or more cells of a tissue, and provoked by the metabolic activation of procarcinogens by phase I enzymes such as P-450 monoxygenases to form ultimate carcinogens responsible for DNA mutations.

Promotion involves colonal proliferation of the initiated cells, and makes them into premalignant tumor cells. Progression is the stage converting such benign tumor cells into malignant tumors. Our attentions have been directed to the inhibition of the tumor promotion stage as a chemopreventive strategy (2) since it needs a long time to occur and is recognized as a reversible process, at least in its earlier stages, among the multistages of carcinogenesis. Extensive epidemiological surveys have demonstrated that ingestion of vegetables or fruits is inversely related to cancer incidence. In fact, citrus fruits are widely known to contain various types of chemopreventers such ad d-limonene, limonoids and their glucosides, flavonoids, carotenoids, and so forth (3–6). In this article, we describe the isolation and identification of a new-type of citrus derived chemopreventer from natsumikan (*Citrus natsudaidai* HAYATA), and the cancer preventive activity of the major compound, auraptene, in mouse skin and rat colon as well as the possible action mechanism so far investigated. The occurrence and quantity of the active constituent (auraptene) in several citrus fruits and their commercially available juices are also reported.

Isolation of the Active Constituents from Natsumikan

To more efficiently search for the active compounds from crude extracts, we have been using a convenient *in vitro* assay, the tumor promoter-induced Epstein-Barr virus (EBV) activation test in Raji cells (7–9). The whole parts of fresh natsumikan were extracted by an FMC in-line Citrus Juice Extractor, which is used in fruit juice factories, to give cold-pressed oil. The oil was then fractionated by silica gel column chromatography with reference to the results of EBV assay. The EBV activation ;inhibitors were finally purified by HPLC to give five coumarin- and psoralen-types of compounds (Fig. 1); isoimpreatorin (IC_{50} = 17 μM), auraptene (IC_{50} = 18 μM), epoxyauraptene (IC_{50} = 27 μM), 5[3,7-dimethyl-6-epoxy-2-octenyl)oxy]psoralen (IC_{50} = 15 μM), and umbelliferone (IC_{50} = 450 μM)(10).

Anti-tumor Promoting Activity in Mouse Skin

Auraptene was then subjected to *in vivo* experiments on account of sample availability. In a two-stage carcinogenesis experiment with 0.19 μmol of 7,12-dimethylbenz[a]anthracene (DMBA) and 1.6 nmol of 12-O-tetracanoylphorbol-13-acetate (TPA) in ICR mouse skin, topical application of 160 nmol auraptene 40 min prior to each TPA treatment significantly reduced tumor incidence and the numbers of tumors per mouse by 27% (P < 0.01) and 23% (P < 0.05), respectively, at 20 weeks after promotion (Table 1). Nishino et al. previously reported anti-tumor promoting activity of a coumarin-related mixture, Pd-II[(+) anomalin, (+) praeuptorin B], in mouse skin treated with DMBA and TPA (11). The present study supports their conclusions that coumarins are potential agents for anti-tumor promotion.

Inhibitory Effects on Reactive Oxygen Wpecies Generation *in vitro*

We examined inhibitory effects of auraptene on TPA-induced superoxide (O_2^-) and hydroperoxide (ROOH) generation in a promyelocytic leukemia cell line, HL-60 cells, differentiated by the treatment of DMSO (12), and on lypopolysaccharide (LPS)/interferon-γ)-induced nitric oxide (NO) generation in a murine macrophage cell line, RAW 264.7 cells (13). Auraptene at a concentration of 10 μM inhibited O_2^- generation by 89% (Table 1). Having no O_2^- scavenging effect (data not shown), auraptene might suppress the multicomponent NADPH oxidase system

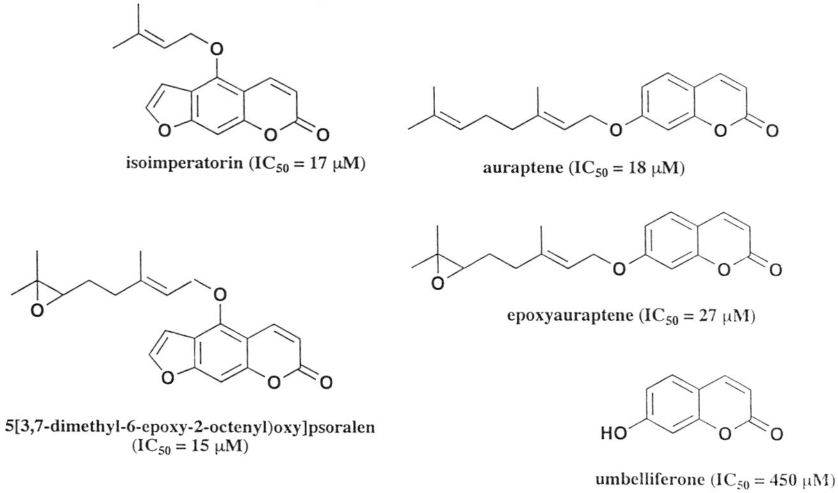

isoimperatorin (IC$_{50}$ = 17 μM)

auraptene (IC$_{50}$ = 18 μM)

epoxyauraptene (IC$_{50}$ = 27 μM)

5[3,7-dimethyl-6-epoxy-2-octenyl)oxy]psoralen
(IC$_{50}$ = 15 μM)

umbelliferone (IC$_{50}$ = 450 μM)

Fig. 1.   Structures of psoralens and coumarins isolated from natsumikan.
Values in parenthesis indicate IC$_{50}$ values of the inhibitory test of EBV
activation

Table 1. *In vivo* and *in vitro* anti-tumor promoting activity of auraptene

	% inhibition (dose or concentration)
In vivo[a]	
no. tumors/mouse	27 (160 nmol, P < 0.01)
	19 (16 nmol, P < 0.01)
tumor incidence	23 (160 nmol, P < 0.05)
	< 10 (16 nmol, not significant)
In vitro	
superoxide[b]	89 (10 μM)
hydroperoxide[c]	83 (10 μM)
nitric oxide[d]	88 (10 μM)

[a]Each group was composed of 15 female ICR mice. Mice at 7 weeks old were initiated with DMBA (0.19 μmol). One week after initiation, the mice were promoted with TPA (1.6 nmol) twice a week for 20 weeks.

[b]HL-60 cells were preincubated with 1.25% DMSO at 37 °C for 4 days, differentiating them into granulocyte-like cells. TPA (100 nM)-induced extracellular O_2^- was detected by the cytochrome c reduction method, in which visible absorption at 550 nm was measured.

[c]TPA (100 nM)-induced intracellular ROOHs in differentiated HL-60 cells were detected by using DCFH-DA as an intracellular fluorescence probe.

[d]LPS (100 ng/mL)/IFN-γ(100 U/mL)-induced NO generation in RAW 264.7 cells was detected by the Griess method, in which visible absorption at 543 nm was measured.

responsible for O_2^- generation. The flow cytometric analyses of differentiated HL-60 cells were then conducted to evaluate suppressive activity toward the TPA-induced intracellular ROOH formation using dichrolofluorescene diacetate as a probe (*14*). Auraptene at a concentration of 10 μM inhibited ROOH formation by 83% (Table 1). Decrease in the ROOH level should be derived from O_2^- generation suppression to some extent. The action mechanism by which auraptene inhibits tumor promotion has not yet been clarified. However, suppression of TPA-induced oxidative stress is presumably involved in the inhibition since free radicals are relevant to tumor promotion in the mouse skin model (*15*). In this model, a great number of leukocytes such as neutrophils, recruited by chemotactic factors, are known to accumulate in the dermis to generate O_2^- with the second TPA stimulation through the NADPH oxidase system (*15*). Interestingly, Petruska et al., recently reported that reactive oxygen species, produced by myeloperoxidase in leukocytes, enhance formation of (±)-trans-7,8-dihydroxy-7,8-dihydrobenzo[a]apyrene[DNA adducts in lung tissue *in vitro* (*16*). it is, thus, tempting to speculate that auraptene is protective to DNA mutations.

furthermore, 50 mM auraptene completely inhibited the LPS/IFN-g-induced NO generation in RAW cells by 93%, and showed no NO scavenging activity (data not shown). Recently Nishino et al., reported that some of the NO-donors have tumor promoting activity in Sencar mouse skin, and tumor promoters induce NO generation in cultured mouse peritoneal macrophages (*17*) and rat hepatocytes (*18*). In addition to its relevance to tumor promotion, NO is considered to have tumor initiating activity due to the formation of nitrosamines (*19*). Thus, reduction of an excess amount of NO would be a novel and reliable means of chemoprevention. In this context, it should be noted that auraptene blocks both O_2^- and NO generating pathways in leukocytes.

Inhibition of Rat Colonic Aberant Crypt Foci Formation

Cancer preventive efficacy of auraptene in the initiation phase was examined in the azoxymethane (AOX)-induced aberrant cyrpt foci (AXF) formation experiments in rat colon. ACF are considered to be preneoplastic lesions since they appear in carcinogen-treated rodent colons and even in human colons with a high risk for cancer development (*20*). In fact, most inhibitors of ACF formation have been proven to possess anticarcinogenesis activity in long-term experiments (*21*). The experimental protocol and results are shown in Fig.2. Auraptene at a dose of 100 or 500 ppm in the basal diet clearly, and dose-dependently, inhibited the number of ACF per colon [inhibitory rate (IR) = 40.8–56.1%)], the number of total AC (IR = 50.8–65.9%), and mean crypt multiplicity (IR = 18.6–24%) after five weeks of feeding (Fig. 2). On the basis of these results, long-term animal experiments for rat colon anti-carcinogenesis is now in progress.

Glutathione S-transferase-inducing Activity in Mouse Liver

Coumarin-related compounds have primarily received attention for the anti-carcinogenic effects in rodents on account of their inducing activity of phase II enzymes such as glutathione S-transferase (GST, EC 2.5.1.18), and their conjugation with electrophilic forms of carcinogenic metabolites. The GST-inducing activity of auraptene in the mouse liver was measured by the method described by Habig et al. using 1-chloro-2,4-dinitrobenzene (CDNB) as a substrate (*22*). As shown in Fig. 3, oral administration of auraptene (50-200 mg/kg body wt.) once a day for five days to male ddy mice enhanced dose-dependent liver GST activity by 1.4-2.1-fold without significant differences in body weight (data not shown). The activity was comparable to that of 3-tert-butyl-4-hydroanisole (BHA). Thus, the GST-inducing activity of auraptene may play an important role in the inhibition of rat colonic ACF formation.

Quantitative Analyses of Auraptene in Citrus Fruits and Their Juices

Levels of auraptene were high (408-585 mg/kg fresh wt.) in the peels of natsumikan and hassaku orange, moderate (101-120 mg/kg fresh wt.) in grapefruit, and absent (< 1 mg/kg fresh wt.) in the peels of Satsuma mandarin (tangerine), Valencia orange, navel orange, lemon, and lime (data not shown). The tendency of auraptene content in the sarcocarps of the above fruits is similar with that in the peels (Fig. 4). The net contents in the sarcocarps, however, were about 40-1,800 times less than those in the peels. Commercial juices from natsumikan, hassaku orange, or grapefruit showed

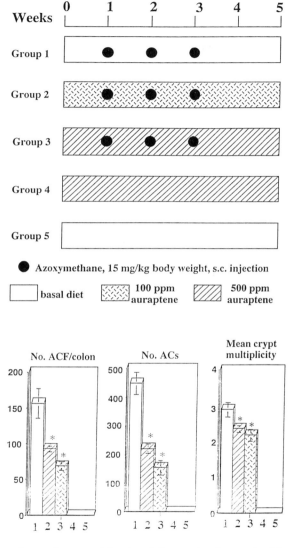

Fig. 2.   Experimental protocol (top) and the results (bottom) of the inhibitory test of azoxynethane-induced aberrant crypt foci (ACF) formation in F344 rat colons.
*P < 0.001.

92

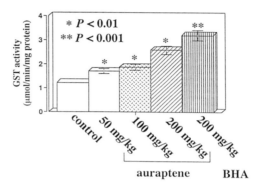

Fig. 3. Effects of auraptene and BHA on glutathione S-transferase activity in ddy mouse liver.
*, **Significantly different from control. Each column represents the mean ± S.E. (n = 5-6). Administration; p.o. (1 time/day for 5 days)

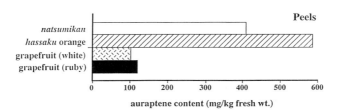

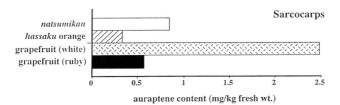

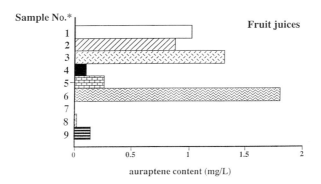

Fig. 4. The results of quantitative analyses of auraptene in the peels (top), sarcocarps (middle), and fruit juices collected in Japan (bottom).
[1]Natsumikan + hassaku orange (fruit juice content: FJC = 100%); [2]hassaku orange (FJC = 100%); [3]hassaku orange (FJC = 50%); [4]Natsumikan + hassaku orange + iyokan (FJC = 10%); [5]Natsumikan + hassaku orange (FJC = 10%); [6]grapefruit (FJC = 100%); [7]Valencia orange (FJC = 100%); [8]Valencia orange (FJC = 100%); [9]Satsuma mandarin (tangerine)(FJC = 100%)

94

relatively higher contents of auraptene (0.87–1.80 mg/L). However, even if hassaku orange or natsumikan is added to the juices, the auraptene content is trivial (0.1–0.26 mg/L) in the products of which total fruit juice content was 10% (Fig. 4).

Conclusions

Auraptene, a citrus coumarin, inhibited mouse skin tumor promotion and rat colonic ACF formation, suggesting that auraptene is an effective cancer chemopreventer. One possible action mechanism for the inhibition of tumor promotion might be suppression of the generation of O_2^-, ROOH, and NO in leukocytes. GST-inducing activity may contribute to the inhibition of ACF formation. As the cold-pressed oils of citrus fruits are produced world wide at the industrial level today, samples of auraptene are readily available. These characteristics together with high chemopreventive potency make it an appropriate source substance for the creation of physiologically functional foods.

Acknowledgments

This study was partly supported by a Grant-in-Aid from the Ministry of Health and Welfare for the Second Term Comprehensive 10-Year Strategy for Cancer Control, Japan, Program for Promotion of Basic Research Activities for Innovative biosciences, and the Japanese Research and Development Association for New Food Creation.

Literature cited

1. Pitot, H. C.; Dragan, Y. P. *FASEB J*. **1991,** *5,* 2280-2286.
2. Murakami, A.; Ohigashi, H.; Koshimizu, K. *Biosci. Biotech. Biochem.* **1996,** *60,* 1-8.
3. Lam, L. K. T.; Zhang, J.; Hasegawa, S.; Schut, H. A. J. Inhibition of chemically induced carcinogenesis by citrus limonoids. In *Food Phytochemicals for Cancer Prevention I;* Huang, M. T., Osawa, T., Ho, C.-T., Rosen, R. T.; Eds., ACS Symposium Series 546, 1994, pp 209-219.
4. Miller, E. G.; Gonzales-Sanders, A. P.; Couvillon, A. M.; Hasegawa, S.; Lam, L. K. T.; Sunahara, G. I. Inhibition of oral carcinogenesis by green coffee beans and limonoid glucosides. In *Food Phytochemicals for Cancer Prevention I;* Huang, M. T., Osawa, T., Ho, C.-T., Rosen, R. T.; Eds., ACS Symposium Series 546, 1994, pp 220-229.
5. Crowell, P. L.; Kenna, W. S.; Haag, J. D.; Ahmad, S.; Vedejs, E.; Gould, M. N. *Carcinogenesis* **1992,** *13,* 1261-1264.
6. Tanaka, T.; Makita, H.; Ohnishi, M.; Hirose, Y.; Wang, A.; Mori, H.; Satoh, K.; Hara, A.; Ogawa, H. *Cancer Res.* **1994,** *54,* 4653-4659.
7. Murakami, A.; Nakamura, Y.; Koshimizu, K.; Ohigashi, H. *J. Agric. Food Chem.* **1995,** *43,* 2779-2783.
8. Nakamura, Y.; Murakami, A.; Koshimizu, K.; Ohigashi, H. *Biosci. Biotech. Biochem.* **1996,** *60,* 1028-1030.
9. Murakami, A.; Ohmura, S.; Nakamura, Y.; Koshimizu, K.; Ohigashi, H. *Oncology,* **1996,** *53,* 386-391.
10. Murakami, A.; Kuki, W.; Takahashi, Y.; Yonei, H.; Nakamura, Y.; Ohto, Y.; Ohigashi, H. Koshimizu, K. *Jpn. J. Cancer Res.* **1997,** *88,* in press.

11. Nishino, H.; Okuyama, T.; Takata, M.; Shibata, S.; Tokuda, H.; Tahayasu, J.; Hasegawa, T.; Nishino, A.; Ueyama, H.; Iwashima, A. *Carcinogenesis* **1990,** 11, 1557-1561.

12. Markert, M.; Andrews, P. C.; Babior, B. M. *Methods Enzymol.* **1984,** *105,* 358-365.

13. Ohata, T.; Fukada, K.; Takahashi, M.; Sugimura, T.; Wakabayashi, K. Jpn. *J. Cancer Res.* **1997,** *88,* 234-237.

14. Bass, D. A.; Parce, J. W.; Dechatelet, L. R.; Szejda, P.; Seeds, M. C.; Thomas, M. *J. Immunol.* **1983,** *130,* 1910-1917.

15. Kensler, T. W.; Egner, P. A.; Taffe, B. G.; Trush, M. A. Role of free radicals in tumour promotion and progression. In *Skin Carcinogenesis;* Slaga, T. J., Klein-Szanto, A. J. T., Boutwell, R. K., Stevenson, D. E., Spitzer, H. L., D'Motto, B, Eds.; Alan R. Liss, Inc. New York, Progress in Clinical and Biological Research Vol. 298, 1989, pp 233-248.

16. Petruska, J. M.; Mosebrook, D. R.; Jakob, G. J.; Trush, M. A. *Carcinogenesis* **1992,** *13,* 1075-1081.

17. Hortelano, S.; Genaro, A. M.; Bosca, L. *FEBS Lett.* **1993,** *320,* 135-139.

18. Hortelano, S.; Genaro, A. M.; Bosca, L. *J. Biol. Chem.* **1992,** *267,* 24937-24940.

19. Liu. R. H.; Jacob, J. R.; Hotchkiss, J. H.; Tennant, B. C. *Carcinogenesis* **1993,** *14,* 1609-1613.

20. Bird, R. P. *Cancer Lett.* **1995,** *93,* 55-71.

21. Tanaka, T.; Mori, H. *J. Toxicol. Pathol.* **1996,** *9,* 139-149.

22. habig, W. H.; Pabst, J.; Jakoby, W. B. *J. Biol. Chem.* **1974,** *249,* 7130-7139.

Chapter 9

Anticarcinogenic and/or Antimetastatic Action of Apple Pectin in Experimental Rat Colon Carcinogenesis and on Hepatic Metastasis Rat Model

K. Tazawa[1,2], Hideo Ohkami[2], Iwao Yamashita[2], Yasuharu Ohnishi[2], Tomohiro Saito[2], Masahiro Okamoto[2], Kiichi Masuyama[2], Kazumaro Yamazaki[2], Shigeru Takemori[2], Mitukazu Saito[2], and Hediki Arai[2]

[1]School of Nursing, and [2]Second Department of Surgery, Toyama Medical and Pharmaceutical University, 2630 Sugitani, Toyama City, Toyama 930-01, Japan

Among pectin, apple pectin (AP) exerts a bacteriostatical action on *Staphylococcus aureus, Streptococcus faecalis, Pseudomonas aeruginosa* and *Escherichia coli*. So, it is thought that AP may change the composition of the intestinal flora. In this study, we used water-soluble methoxylated pectin from apples. Supplementation of the diet with 20% AP significantly decreased the number, and the incidence of azoxymethane-induced colonic tumors. Prostaglandin E2 levels in distal colonic mucosa, or portal blood, from 20% AP fed rats were lower than those in basal diet fed rats. Furthermore, 20% AP fed rats significantly inhibited the incidence of the experimental portal bacteremia induced by methotrexate. The incidence of hepatic metastasis was significantly reduced by the diet with 20% AP. In conclusion, AP may exert an antitumor effect, and prevents cancer metastasis and portal bacteremia by modifying host immune function, and/or by altering the intestinal flora. Dietary fibers, like pectin, have a very important function in the intestinal tract as anti-inflammatory foods.

Epidemiologic and animal model studies suggest a protective role of certain types of dietary fiber, against colon cancer. This protection by dietary fiber may be mediated through the dilution of tumorigenic compounds to the gut, and an indirect effect on the metabolism of carcinogens. Experiments were conducted in animal models to study the effect of certain nonconventional, or conventional types of dietary fiber on colon carcinogenesis. Regarding pectin, there are contradictory reports as to whether or not it is effective in experimental colon carcinogenesis (*1*)~(*8*). In fact, there are several kinds of pectin, and the previous investigators mainly used citrus pectin.

Previously, we reported that apple pectin inhibited azoxymethane(AOM)-induced colon carcinogenesis in rats (3). Apple pectin has stronger bacteriostatic action on *Staphylococcus aureus, Streptococcus faecalis, Pseudomonas aeruginosa,* and *Escherichia coli* than do either citrus pectin or carboxymethylcellulose Na (9). Therefore, apple pectin may be expected to have a strong influence on the intestinal microflora and bacterial enzymes activities. The intestinal bacteria may play a significant role in the pathogenesis of colon cancer because their enzymes are important in the metabolism of procarcinogens, and the production of tumor promoters in the colon (10).

To date, there have been no reports concerning the effects of apple pectin on fecal enzyme activities, or on prostaglandin E2(PGE2) levels in colonic mucosa or portal blood, in relation to prevention of experimental colon carcinogenesis. The ability of apple pectin to decrease PGE2 was dose-dependent (11), and those results suggest an anti-inflammatory effect in the bowel.The purpose of the present study was to investigate the effects of dietary apple pectin on three fecal enzymes: β-glucuronidase, β-glucosidase and tryptophanase, at various periods of time. This study also examined the effect of pectin on the portal bacteremia and hepatic metastasis rat model, to determine whether the anti-inflammatory effect of dietary fiber had a scavenger effect in the intestinal digestion and portal circulation system.

Materials and Methods

1) Animals and diets. Male Donryu rats from Shizuoka Laboratory Animal Center (Shizuoka, Japan), 4 weeks old, each weighing about 200g, were used and placed in steel cages to prevent coprophagia. A total of 62 rats were divided into 3 groups; the control group fed a basal diet, the citrus pectin (CP) group fed the basal diet containing 20% CP (USA-SAG type, DD slow set, The Copenhagen Pectin Factory Ltd., Copenhagen, Denmark), and the apple pectin (AP) group fed the basal diet containing 20% AP (OM type, Herbstreith Äï Fox, Neuenburg, Germany). The molecular structure of two pectins is somewhat complicated. AP and CP is formed with D-galacturonic acid molecules which are linked to each other by ÉØ-(1-4)-glycosidic bonds to become polygalacturonic acid. The carboxyl groups are partially esterified with methanol, and the secondary alcohol groups may be partially acetylated.

2) Carcinogenesis. From 2 weeks after starting diet administration, animals (6 weeks old) were given subcutaneous injections of AOM (7.4 mg/kg,Sigma Chemical Company, St. Louis, USA) once weekly for 10 weeks. Rats were sacrificed 30 weeks after the first injection of AOM.

3) Fecal bacterial enzyme activities. Fresh fecal pellets were collected for the following enzyme activities at the 0 (after 2weeks diets), 12 and 19weeks. We previously described the methods of β-Glucosidase, β-Glucuronidase and Tryptophanase assay (*12*).

4) Prostaglandin E2 (PGE2) levels. When rats were sacrificed 30 weeks after the first injection of AOM, the colonic mucosa was scraped off, immediately frozen, and stored at -80Åé until extraction for the measurement of PGE2. The amount of PGE2 in the extract of homogenized colonic mucosa and portal blood was measured by a radioimmunoassay method at Minase Research Institute (Ono Pharmaceutical Co., Ltd. Japan)

5) Portal bacteremia. 3 days after methotrexate (MTX;20mg/kg) was administered into the abdominal cavity, all animals were anesthetized, and phlebotomized, via the portal vein, under strictly aseptic conditions. Blood cultures were then obtained. We examined the incidence of bacteremia between AP group and control group fed a basal diet.

6) Measurement of experimental hepatic metastasis. Under pentobarbital anesthesia, Donryu rats underwent a laparotomy. 2.5 X 10^6 AH60C ascites liver cancer cells were then inoculated into the portal vein(*13*). At 14 days after tumor transplantation, the rats were sacrificed and metastatic nodules in the liver were macroscopically counted.

Results

1) Tumor induction (Table I). The incidence of total colon tumors in the control group and CP group was 100% (18/18) and 70% (14/20) ($P<0.05$), respectively. That of 43% in the AP group (6/14) was significantly lower than that in the control group ($P<0.001$). The average number of tumors per rat in the AP group (0.6±0.2) had a statistically significant difference compared with that in the control group (2.4±0.2) ($P<0.001$).

2) Fecal enzyme activities (Table II)
β-glucuronidase activity: At week 0 (2 weeks after feeding the diet), a significant decrease of activity was seen in the apple pectin group, compared with that in the control group. The activity in the apple pectin group was 1/10 lower than that in the control group. However, from the 12th week on, the activity increased in both pectin diets compared with that in the control diets ($P<0.05$ or less).
Fecal β-glucosidase activity: From week 0, É¿-glucosidase activity tended to decrease in citrus pectin group ($P<0.1$) compared with that in the control group. The apple pectin group, on the other hand, had a significantly lower β-glucosidase activity ($P<0.05$ or less) than did the control diet group.
Fecal tryptophanase activity: In the citrus pectin group, tryptophanase activity decreased, compared with that in the control group at week 0 but there was no tendency. In the apple pectin group, the activity tended to decrease compared with that in the control group.

Table I. Number and Incidence of Colon Tumors in Each Experimental Group

Diet	Animals with colon tumor (%)	No. of tumor per rat
Control (n=18)	18 (100)	$2.4 \pm 0.2^{a)}$
Citrus pectin (n=20)	14 (70)[b)]	$1.6 \pm 0.4^{b)}$
Apple pectin (n=14)	6 (43)[c)]	$0.6 \pm 0.2^{c)}$

a) mean ±SE, b) P<0.05, c) P<0.001 compared with control group

Table II. Fecal Enzyme Activities on the 0, 12th and 19th Week after the First Azoxymethane Injection

Group (n)	0 week (after 2 week diets)	12weeks	19weeks
β-glucuronidase (m mol./mln./g)			
Control (5)	$2.47\pm0.94^{a)}$	1.58 ± 0.28	1.21 ± 0.20
20% CP (5)	1.10 ± 0.22	$3.11\pm0.27^{c)}$	$3.72\pm0.66^{b)}$
20% AP (5)	$0.23\pm0.03^{b)}$	$4.06\pm0.33^{d)}$	$2.56\pm0.41^{b)}$
β-glucosidase (m mol./mln./g)			
Control (5)	2.50 ± 0.46	1.88 ± 0.22	0.92 ± 0.16
20% CP (5)	1.40 ± 0.10	1.34 ± 0.05	1.31 ± 0.27
20% AP (5)	$0.73\pm0.06^{b)}$	$4.06\pm0.33^{c)}$	2.56 ± 0.41
Tryptophanase (n mol./mln./g)			
Control (5)	96.6 ± 21.7	67.0 ± 8.9	49.1 ± 6.5
20% CP (5)	44.4 ± 4.8	64.7 ± 8.4	64.7 ± 5.9
20% AP (5)	68.9 ± 18.7	$39.5\pm6.6^{b)}$	40.6 ± 4.2

a) mean ±SE, b) P<0.05, c) P<0.01, d) P<0.005 compared with control group

3) PGE2 levels (Table III)

The PGE2 level of portal blood in 20% AP-fed rats (0.30Å}0.08 ng/ml) was significantly lower than in the control group (0.81Å}0.17 ng/ml) ($P<0.05$). There was significant difference on PGE2 level in distal colonic mucosa between control group (422.±125.6 ng/g), and 20% AP group (166.6±25.8 ng/g) ($P<0.001$).

4) Portal bacteremia (Table IV)

The incidence of portal bacteremia decreased to 40% in 20% AP-fed rats. Cultured portal blood after MTX injection yielded bacteremia, consisting of predominantly *Escherichia coli* and *Enterococcus*.

5) Hepatic metastasis (Table V)

The hepatic metastasis of rate model was significantly lower in the AP group than in the control group ($P<0.05$). The mean number of tumor nodules per rat in the AP group was significantly lower than that of control group ($P<0.01$).

Discussion

Pectin is a partially methoxylated polymer of galacturonic acid obtained from fruit. There have been a number of reports concerning the effectiveness of pectin in inhibiting experimental colon carcinogenesis (1),(2),(5). There are several kinds of pectin, but these research groups mostly used citrus pectin. Apple pectin has stronger bacteriostatic action against pathogenic bacteria than citrus pectin (9). Therefore, apple pectin may markedly affect the composition of the intestinal bacterial flora. The results of our study indicates that the induction of colon neoplasia by AOM was dose-dependently inhibited by AP. Fecal tryptophanase activity tended to decrease in the AP group compared with that in the control group. The reduced level of tryptophan metabolites in the colon might be related to the inhibitory effect of AP on colon carcinogenesis (Chung).

At week 0 (2 weeks after starting the diet but before AOM), the B-glucuronidase activity in the AP group had fallen to 1/10 of that the control group. Bauer et al. (2) found an increased incidence of dimethylhydrazine-induced colorectal tumors and fecal B-glucuronidase activity in male Sprague-Dawley rats fed a diet containing 6.5% pectin. In general, the higher fecal B-glucuronidase activity in the pectin-fed animals is said to be associated with a higher tumor incidence. In a recent study, we examined B-glucuronidase activity weekly during the initiation phase, and found that the initially lower B-glucuronidase activity in the AP group gradually increased to reach that in the control group at 6 week after starting the diet (data not shown).

It is known that PGE2 has a role in the regulation of the immune response, directly or indirectly, and in the activation of ornithine decarboxylase (ODC), which is necessary for the proliferation of tumors. These results thus indicate that the effect of AP on colon carcinogenesis may partially depend on the decrease of PGE2

Table III. Prostaglandin E2 Levels in Colonic Mucosa and Portal Blood

20% citrus pectin or 20% apple pectin

Group (n)	PGE2 level （ng/g)	
	proximal colon (mean ±SE)	distal colon (mean ±SE)
Control (5)	397.6 ± 62.3	422.1 ± 125.6
20% CP (5)	379.4 ± 77.4	324.9 ± 33.7 [a]
20% AP (5)	274.3 ± 80.6	166.6 ± 25.8 [a]

Apple pectin (10% or 20%)

Group (n)	PGE2 level （ng/ml) portal blood (mean ±SE)
Control (7)	0.81 ± 0.17
10% AP (6)	0.54 ± 0.13
20% AP (7)	0.30 ± 0.08 [b]

a) $P<0.001$, b) $P<0.05$ compared with control group

Table IV. Effect of 20% Apple Pectin-supplemented Basal Diet on the Incidence of Portal Bacteremia 3 Days after Methotrexate Administration

	Control	20% apple pectin
Bacterial culture	4 / 4	2 / 5

Table V. Effect of 20% Apple Pectin on Hepatic Metastasis Produced by Intra-portal AH60C Cells

Group	Hepatic metastasis	
	incidence	Av. no. of tumor nodules / rat
Control	14 / 15 (93.3%)	56.3 ± 12.5 [a]
20% AP	7 / 13 (53.9%) [b]	16.2 ± 5.4 [c]

a) mean ± SE, b) $P<0.05$, c) $P<0.01$ compared with control group

concentration in the colonic mucosa. GE2 levels in the blood of the portal vein in the 20% AP group were significantly lower on the 30th week than PGE2 levels in the control group. Also Oral administration of 20% AP inhibited the portal bacteremia induced with MTX and hepatic metastasis of AH60C asites liver cancer cells inoculated into the portal vein.

These results suggest that AP has a scavenger effect in intestinal digestion. The detailed mechanisms of action of AP have not yet been clarified. However, findings obtained so far, including the results of the present study, suggest that AP exerts an antitumor effect and prevents cancer metastasis and carcinogenesis by modifying host immune function, and/or by altering the intestinal flora. The inhibition of hepatic metastasis by oral administration of AP in the present study strongly suggests that this dietary fiber may be effective for the prevention of micrometastasis, and/or residual cancer cells remaining after surgery.

Various drugs and foods have been suggested to have a preventive effect on cancer, but few of them have proven to be clinically effective. Preparation of galacturonic acid obtained from apple may be appropriate for prophylactic use because they have few or no adverse effects. Although much labor and time is needed to demonstrate the benefit of medical interventions for disease prevention, it is hoped that investigation of the prevention of cancer metastasis using AP and other anti-inflammatory food preparations will make considerable progress in the future.

Conclusion

AP administered orally has a scavenger effect in the intestinal digestion and portal circulation system. Dietary fibers like pectin, espescially apple pectin with strong bacteriostatic action, have a very important function in the intestinal tract as anti-inflammatory foods.

REFERENCES

(1) Watanabe,K.; Reddy,B.S.; Weisburger J.H.; Kritchevsky,D.J. Natl. Cancer Inst.**1979**,63,141.
(2) Bauer,H.G.;Asp,N.G.;Dahlqvist,A.;Fredlund,P.E.;Nyman,M.;Oste,R. Cancer Res. **1981**, 41, 2518.
(3)Tazawa,K.;Okami,h.;Yamashita,I.;Shimizu,T.;Fujimaki,M.;Murai,K.;Kobashi,K.; Honda,T.In Recent Advantage in Management of Digestive Cancers; Takahashi,T.; Springer-Verlag,Tokyo,Japan,**1993**,Vol.1.pp471-473.
(4)Golden,B.R.;Gorbach,S.L. J. Natl. Cancer Inst.**1978**, 57, 371.
(5)Freeman,H.J.;Spiller,G.A.;Kim,Y.S. Cancer Res. **1980**, 40, 2661.
(6)Masaki,K. J. Wakayama Med. Soc. **1993**, 44, 351.
(7)Freeman,H.J. Cancer Res. **1986**, 46, 5529.
(8)Prizont,R. Cancer Res. **1984**, 44, 557.
(9)Tazawa,K. Proceedings of the 7th Biennial Congress of World Council of Enterostomal Therapists; Perth, Australia,**1988**,Vol.1,pp37-41.

(10)Ohkami,H.;Tazawa,K.;Yamashita,I.;Ohnishi,Y.;Kobashi,K.;Fujimaki,M Jpn. J. Cancer Res. **1995**, 86, 523.

(11)Tazawa,K.;Ohkami,H.;Yamashita,I.;Ohnishi,Y.;Kobashi,K.;Fujimaki,M. J. Exp. Clin. Cancer Res. **1997**, 16, 33.

(12)Akao,T.: Kobashi,K. Chem. Pharm. Bull. **1987**, 35, 705.

(13)Tazawa,K.;OkamiH.;Ohnishi,Y.;Saito,T.;Okamoto,M.;Yamamoto,K.;Takemori, S.;Arai,H. Biotherapy **1997**, 11, 524.

Chapter 10

Antitumor Activity of *Emblica officinalis Gaertn* Fruit Extract

Pratima Sur[1], D. K. Ganguly[1], Y. Hara[2], and Y. Matsuo[3]

[1]Division of Pharmacology and Experimental Therapeutics, Indian Institute
of Chemical Biology, 4 Raja S.C. Mullick Road, Jadavpur, Calcutta 700 032, India
[2] Food Research Laboratory, Matsui Norin Company Ltd., 223-1 Miyabara, Fujieda
City, 426 Japan
[3]Hayashibara Biochemical Laboratory, Fujisaki Cell Center, Okayama 702, Japan

The antitumor activity of *Emblica officinalis Gaertn (Phyllanthus
emblica Linn)* fruit extract (EFE) was assessed against different
human leukemic cell lines, ML-2, U-937, and K-562. Significant
cell growth inhibition was observed in all the cell lines. Induction
of differentiation, assessed by indirect immunofluorescence,
Nitroblue tetrazolium reduction, morphology was observed in EFE
treated cells. Balb/C mice, pretreated with the EFE extract,
inoculated with Ehrlich Ascites Carcinoma (EAC) cells showed
significant cell growth inhibition. The *in vivo* effect was
accompanied by the anti-inflammatory property of EFE.

In Indian traditional medicine, the fruit of *Emblica officinalis Gaertn* has been used
extensively as the main ingredient, which has been known to prevent colds, coughs,
and enhances immunity (1). So far, the fruit is known as one of the best sources of
natural Vitamin C. Such Vitamin C has been found to be more readily assimilated
than the synthetic Vitamin C. The fruit is also known to contain a significant amount
of pectin (1), a complex polysaccharide containing galactoside residues, which is
known to possess anticancer properties (2). The antioxidant, and strong reducing
properties of Vitamin C are known to be responsible for its use as a free radical
scavenger, indicating its chemotherapeutic potential (3). Many anti-inflammatory
drugs are reported to prolong survival of patients with cancer (4). The anti-
inflammatory property of this fruit is also mentioned in Indian traditional medicine.
Considering all previous facts, the present study was undertaken using EFE extract.
The *in vitro* antitumor property against human leukemic cell lines, preliminary *in
vivo* antitumor property against Ehrlich Ascites Carcinoma (EAC), and the anti-
inflammatory property of EFE were all evaluated in this study.

Plant Material. The fruit of *Emblica officinalis* was collected during the winter
season (November-December) in West Bengal, India. Voucher specimens of the

dried fruit and extract were deposited at the Division of Pharmacology and Experimental Therapeutics, Indian Institute of Chemical Biology, Calcutta, India.

Preparation of *Emblica officinalis* Fruit Extract (EFE). The fruit is green when young, and gradually turns to pale yellow when it is mature. The EFE was prepared in the following manner:

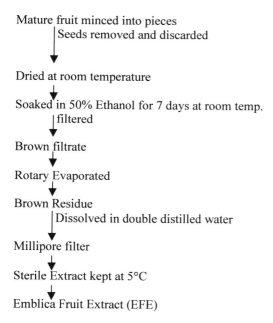

Mature fruit minced into pieces
Seeds removed and discarded

Dried at room temperature

Soaked in 50% Ethanol for 7 days at room temp.
filtered

Brown filtrate

Rotary Evaporated

Brown Residue
Dissolved in double distilled water

Millipore filter

Sterile Extract kept at 5°C

Emblica Fruit Extract (EFE)

Cells. Human leukemic cell lines ML-2, U-937, and K-562 cells were obtained from the Fujisaki Cell Center, Okayama, Japan. Cells were maintained routinely in a RPMI 1640 medium, and supplemented with 10% fetal calf serum (heat inactivated). Cultures were maintained at 37°C in a 95% humidified atmosphere containing 5% CO_2 in air.

For preliminary *in vivo* studies, Ehrlich Ascites Carcinoma (EAC) cells were obtained from the Chittaranjan National Cancer Research Center, Calcutta, India. EAC cells were maintained in Balb/C mice by weekly intraperitoneal inoculation with 10^6 cells per mouse. Sprague Dawley rats (125-150gm) were used to study the anti-inflammatory effect.

***In Vitro* Cell Growth Inhibition.** Cells at a concentration of about 0.5×10^6 /ml were used to start the experiment in a 12-well sterile plastic plate. EFE, at desired concentrations, were added on day '0', and incubated for 5 days at 37°C. Viable cells, as judged by the Trypan blue dye exclusion test, were counted every day (5).

Cell growth inhibitions, after treatment, were indicated by doubling times, using the formula:

Doubling Time = Time(hour)/No. of Doubling
Where, No. of Doubling = [log.final cell no. - log.initial cell no.] /0.302

Tritiated Thymidine, Uridine and Leucine Incorporation. Effect of EFE (50, 100, 200 µg/ml) on the synthesis of DNA, RNA, and protein synthesis were observed with EFE treated ML-2 cells after 24 and 72 hour time intervals. The method used was essentially the same as reported earlier (5).

Cell Cycle Analysis. Cell cycle analysis was performed with EFE treated ML-2 cells as described earlier (5). Treated ML-2 cells, after 24 and 72 hours, were washed with phosphate buffered saline, and fixed in 70% ethanol. The fixed cells were washed and incubated with RNase (40 µg/ml) for 30 min at 37°C. Cells were again washed and stained with propidium iodide (50 µg/ml) for 30 min. at 0°C. The stained cells were analyzed on a coulter profile flow cytometer.

Indirect Immunofluorescence. Cells, after 72 hours of treatment, were washed in 0.9% saline, and incubated for 30 min. at 37°C in heat inactivated, pooled, AB serum to prevent nonspecific FC binding (6). Modulation of the expression of specific cell surface antigens was observed by monoclonal antibodies obtained from Nichirei (Tokyo, Japan), Coultronics and Seralab (UK). Cells were incubated for another 30 min. at 37°C with a second monoclonal antibody containing FITC conjugated (ab')$_2$ goat anti-mouse antiserum. The cells were washed and suspended in 0.9% saline, and analyzed by a coulter profile flow cytometer (Coulter Electronics, Hialeah, F1), as done earlier (7).

Morphology. Morphologic differentiation was observed on Giemsa-stained slide preparations, with 72 hour EFE treated ML-2 cells.

Nitroblue tetrazolium (NBT) reduction assay. The NBT reduction assay was performed, as described previously (7). EFE treated ML-2 cells were washed and suspended (1 x 10^5 cells) in 0.2 ml of RPMI 1640 meduim, containing 5% FCS, 0.1% NBT, and 30 mg of TPA. After 30 minutes of incubation at 37°C, the percentage of cells containing blue black deposits, were counted.

***In vivo* effect of EFE on the growth of EAC in mice.** *In vivo* effects with EFE, against EAC, were observed following three different methodologies:

A Animals (n=7) were pretreated with 100 mg/kg EFE for 3 days. After a 24 hour gap, the animals were inoculated with 10^5 EAC cells intraperitoneally. On the 4th day of inoculation, the animals were sacrificed. Total intraperitoneal EAC cells were then counted, and compared with the control (without treatment) group.

B. Animals (n=7) were inoculated with 10^5 EAC cells/mouse on day zero. Treatment with EFE (100 mg/kg) started from day 1, and continued to day 3. On day 4, the animals were sacrificed. Intraperitoneal cell counts were then made, and compared with the control.

C. Animals (n=7) were pretreated for 3 days as described in A. After a 24 hour period, they were inoculated, and treated as described in B. On day 4, they were sacrificed. Intraperitoneal EAC cells were then counted, and compared with the control.

Antiinflammatory activity (Carrageenan-induced Oedema). A 1% solution (0.1 ml) of carrageenan, in 0.9% saline, was injected in male rats beneath the plantar aponeurosis of the right hand paw, and divided into four groups (6 in each). Group 1 served as the control, group 2 received phenylbutazone (100 mg/kg, ip) as standard. Groups 3 and 4 received 50 mg/kg, ip and 100 mg/kg, ip of EFE respectively. Drugs were administered 1 hour before the injection of carrageenan . Foot volume was measured plethesmographically after 4 hours of carrageenan administration. The difference between the control and the treated groups indicated the degree of oedema developed, and the percentage of inhibition was thereby calculated. For the *in vivo* experiments, significance tests were done using the Student's t-test.

Results

Differential cell growth inhibitions, as indicated from the doubling times, were observed in EFE treated cell lines (Table I). The prolonged doubling time of treated cells with respect to the control, indicate inhibition of cell growth. Negative doubling time was due to continuous cell death after treatment. The EFE treated ML-2 cells showed the maximum sensitivity to EFE. To make a comparitive study, the time required for different cell lines to inhibit 50% cell growth, with a specific dose of EFE (200 μg/ml of cell suspension), was made (Table II). It was found that the ML-2 cell line was the most sensitive, requiring 32 hours to inhibit 50% cell growth. For the K-562 cell line, the time required was 63.0 hours. The ML-2 cell line was selected for further study because of its higher sensitivity against EFE.

Cell growth inhibition of the ML-2 cells was corroborated by the inhibition of ^{3}H-Thymidine uptake after treatment, with different doses of EFE (Table III). With 100 μg EFE/ml of ML-2 cell suspension, 65% inhibition of ^{3}H-Thymidine uptake could be observed in 72 hours. However, uptake of ^{3}H-Uridine or ^{3}H-Leucine was not inhibited in EFE treated ML-2 cells (data not shown). The DNA histogram patterns of EFE treated ML-2 cells were found to be perterbed after the 72 hour treatment (Table IV). Accumulation of cells in the 'G$_2$-M' compartment was prominent with a depletion of 'S' phase cells, resulting in the inhibition of cell growth.

Table I. Doubling Time in Log Phase of Different Tumor Cells under the Effect of EFE

Tumor	Cell Lines	Doubling Time		(Hr)	
		EFE Concentration		(μg/ml)	
		0	100	200	400
1.	ML-2	30.5	38.7	64.8	-58.9
2.	U-937	36.2	39.5	50.2	98.3
3.	K-562	29.6	30.7	43.6	100

Table II. Time Required for 50% Cell Growth Inhibition with EFE

Cell Line	I_{50}	(Hr)
	100 μg/ml	200 μg/ml
ML-2	44.0	32.0
U-937	48.0	40.0
K-562	92.0	63.0

Specific antibodies expressed on macrophage, monocyte, and granulocyte surfaces were observed on the ML-2 cells treated with EFE (100μg/ml) for 72 hours (Table V). Such induction of differentiation, evidenced by indirect immunoflorescence, was supported by enhanced NBT reducing activity (Table VI), and morphological changes (Figure 1). The morphology of different doses of EFE (72 hr) treated cells, showed enhanced cytoplasm to nuclear ratio, and ruffled surfaces. Morphologically mature cells were observed with all the doses of EFE used.

Preliminary *in vivo* cell growth inhibitions are shown in Table VII. Pretreatment with the EFE, followed by i.p. EAC tumor inoculation, showed a 50% tumor growth when compared with the control (protocol-A). When tumor bearing mice were treated with EFE, as described in protocol-B, they produced a 30% inhibition of the tumors. When the protocols were coupled together, the cell growth inhibition was found to be 82%, showing an additive effect. It is important to note that experiments carried out with extracts of EFE, prepared from fruits from different areas of West Bengal, India, showed similar results as those described above. The appreciable antiinflammatory activity of EFE was found when compared with the control. The treatment with 100 mg/kg of phenylbutazone showed 50% inhibition. 40% and 62.3% (p<0.001) inhibitions of oedema were observed with 50 mg/kg and 100 mg/kg EFE treatments, respectively.

Table III. Inhibition of DNA Synthesis

	^{3}H-Thymidine	Incorporation in ML-2	Cell
	24 Hr Treatment	48 Hr Treatment	72 Hr Treatment
EFE μg/ml		c.p.m. /10^5 cells	
0(Control)	18,466(--)	20,503(--)	23,921(--)
50	15,350(17)	14,783(28)	11,728(51)
100	15,542(16)	10,529(48.7)	8,370(65)
200	10,742(42)	9,270(54.8	7,250(70)

The results shown represent the average value of the three independent experiments (deviations were within 3-5%). The values in parenthesis indicate percent inhibitic n.

Table IV. Cell Cycle Analysis
DNA Histogram of ML-2 Cells Treated with EFE

Intensity Distribution of Stain Fluorescence at Different Stages of Cell Growth

EFE mg/ml	48 Hr				72 Hr		
	G_0-G_1	G_2-M	S		G_0-G_1	G_2-M	S
0(Control)	39.9	17.5	42.5		36.25	12.9	50.8
50	34.7	13.0	52.2		33.44	23.45	43.1
100	34.5	17.8	47.7		38.14	20.35	41.5

Discussion.

The purpose of the present study was to investigate the antitumor property of EFE. *In vitro* inhibition of cell growth in ML-2 cells, by EFE, was accompanied by inhibition of DNA synthesis and perterbation of the cell cycle.

Growth and maturation are normally synchronized in hematopoitic stem cells. A blocked cellular maturation plays an important role in cancer. Thus, therapy leading to the differentiation is often used to induce stem cells to mature (8,9). Enhanced expression of myeloid differentiation antigens CD_{14}, CD_{15}, CD_{11b}, etc. are normally expressed on the surface of monocytes, macrophages and activated granulocytes, and were found to be enhanced on the surface of EFE treated ML-2 cells, indicating differentiation.

It has been observed (10,11) that antineoplastic agents, such as inducers of differentiation, cause sustained inhibition of DNA synthesis, while protein or RNA synthesis remains unchanged. Similar observations were found with EFE treated ML-2 cells.

Table V. Indirect Immunofluorescence
Reactivity of ML-2 Cells with Antibodies on Monocyte, Macrophage and Granulocyte Surface Antigens after a 72 Hr. Treatment

Antibody	Cluster Designation	Mean Fluorescence Intensity(MFI)		
		Saline Control	50 µg/ml EFE	100 µg/ml EFE
1. MY 9	CD_{33}	1.592	1.254	1.376
2. MY 4	CD_{14}	0.298	0.312	0.347↑
3. M0	CD_{11}	0.288	0.294	0.346↑
4. Leu-15	CD_{11b}	0.279	0.280	0.317↑
5. MCS-1	CD_{15}	0.300	0.298	0.366↑
6. Saline	--	0.284	0.279	0.315

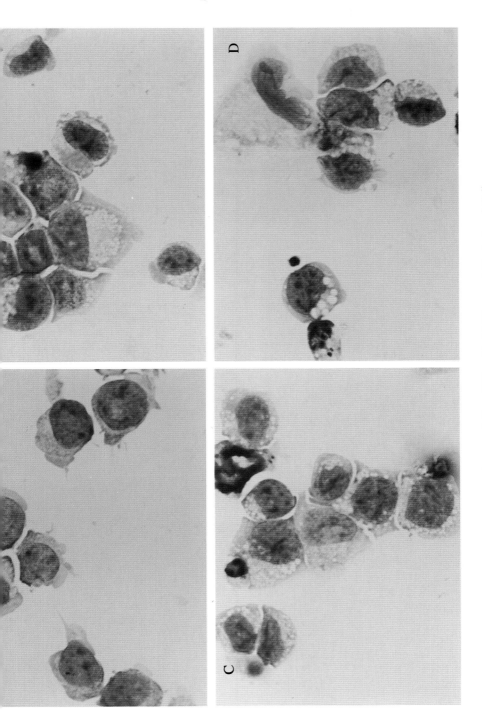

Fig. 1. Cell Morphology of ML-2 Cell at 72Hr with (a) Untreated Control and Cells treated with (b) 50μg/ml, (c) 100μg/ml (d) 200μg/ml of EFE.

**Table VI. Nitroblue Tetrazolium Reduction
NBT-Reducing Activity After EFE Treatment in ML-2 Cells**

Experiment(n=5)	Percent 24 Hr	Positive Cell ±S.D. 48 Hr	72 Hr
Control	0.52 ± 0.0	0.78 ± 0.09	1.2 ± 0.7
EFE 50 µg/ml	1.0 ± 0.5	5.8 ± 1.2	7.5 ± 0.6
EFE 100 µg/ml	5.3 ± 2.1	11.9 ± 2.3	12.0 ± 2.1
EFE 200 µg/ml	6.05 ± 3.1	12.0 ± 2.8	14.5 ± 1.8

Vitamin C in the extract might be partially damaged inspite of every care, due to exposure during the drying of the fruits, and the preparation of EFE. The amount of Vitamin C present in the extract, after drying, possibly acted as a scavenger of free radicals and prevented tumor growth. Pectins, which are complex polysaccharides, and known to inhibit azomethane, induced colon carcinogenesis (12), adenocarcinoma growth, and embolization of tumor cells (13), were reported to be present in the fruit (1). Such properties of Vitamin C and pectins might be responsible for the antitumor property of the fruit extract. Further *in vitro* studies with EFE are in progress with different fractions of EFE.

Table VII. *In Vivo* Cell Growth Inhibition

Treatment[a]	Total Number of Viable Intaperitoneal EAC cell / mouse (x10^7)	Percent Cell Growth Inhibition
A. Pretreatment	$0.72 \pm 0.08*$	50.0
B. Post Treatment	$1.00 \pm 0.2*$	30.0
C. Pretreatment + Post Treatment	$0.25 \pm 0.18*$	82.5
D. Control	1.43 ± 0.14	---

[a]*In vivo* treatment in Balb/C mice with 100 mg/kg EFE, following methodologies as described in the text.
*$p<0.001$ (n=7)

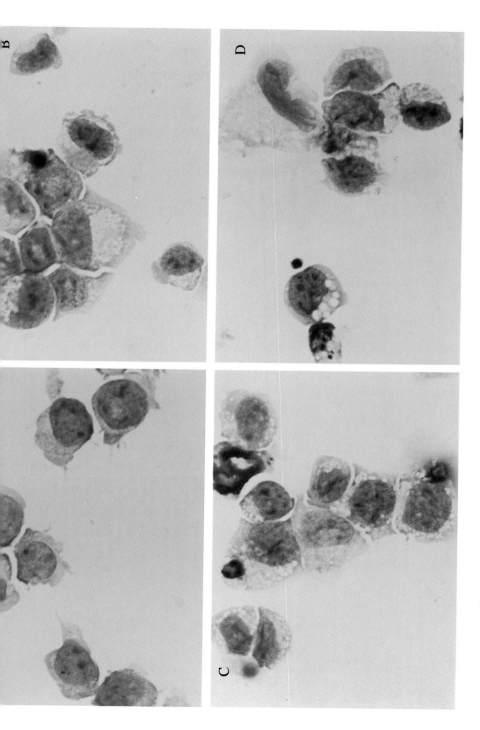

Fig. 1. Cell Morphology of ML-2 Cell at 72Hr with (a) Untreated Control and Cells treated with (b) 50μg/ml, (c) 100μg/ml (d) 200μg/ml of EFE.

**Table VI. Nitroblue Tetrazolium Reduction
NBT-Reducing Activity After EFE Treatment in ML-2 Cells**

Experiment(n=5)	Percent	Positive Cell ±S.D.	
	24 Hr	48 Hr	72 Hr
Control	0.52 ± 0.0	0.78 ± 0.09	1.2 ± 0.7
EFE 50 µg/ml	1.0 ± 0.5	5.8 ± 1.2	7.5 ± 0.6
EFE 100 µg/ml	5.3 ± 2.1	11.9 ± 2.3	12.0 ± 2.1
EFE 200 µg/ml	6.05 ± 3.1	12.0 ± 2.8	14.5 ± 1.8

Vitamin C in the extract might be partially damaged inspite of every care, due to exposure during the drying of the fruits, and the preparation of EFE. The amount of Vitamin C present in the extract, after drying, possibly acted as a scavenger of free radicals and prevented tumor growth. Pectins, which are complex polysaccharides, and known to inhibit azomethane, induced colon carcinogenesis (12), adenocarcinoma growth, and embolization of tumor cells (13), were reported to be present in the fruit (1). Such properties of Vitamin C and pectins might be responsible for the antitumor property of the fruit extract. Further *in vitro* studies with EFE are in progress with different fractions of EFE.

Table VII. *In Vivo* Cell Growth Inhibition

Treatment[a]	Total Number of Viable Intaperitoneal EAC cell / mouse (x10^7)	Percent Cell Growth Inhibition
A. Pretreatment	0.72 ± 0.08*	50.0
B. Post Treatment	1.00 ± 0.2*	30.0
C. Pretreatment + Post Treatment	0.25 ± 0.18*	82.5
D. Control	1.43 ± 0.14	---

[a]*In vivo* treatment in Balb/C mice with 100 mg/kg EFE, following methodologies as described in the text.
*$p < 0.001$ (n=7)

The *in vivo* cell growth inhibitions, according to different protocols, showed the differential efficacy of EFE. Moreover, the enhanced antitumor effect in EFE pretreated mice, followed by EAC inoculation, and indicated the preventive character of the extract. Vitamin C in the fruit extract helped in the destruction of the free radicals responsible for the onset and progression of cancer.

Inflammation is generally considered to be closely related to tumor promotion. It has been shown that anti-inflammatory agents might be able to retard the tumor growth by inactivating the synthesis of prostaglandin, which is related to cell proliferation and neoplasia (14). The strong anti-inflammatory effect of EFE might possibly be responsible, at least partially, for its anti-inflammatory property. Based on the present work, it can be concluded that EFE might be an important candidate in cancer therapy.

Acknowledgments

University Grants Commission, India is acknowledged for Research Scientistship to P. Sur. Thanks are due to Prof. Takayuki Shibamoto, University of California, Davis for fruitful discussions. The authors acknowledge Dr. Pallab Seth, Division of Pharmacology, Indian Institute of Chemical Biology, Calcutta for helping in the preparation of the manuscript, and Mr. R.M. Sharma of the same division for excellent technical assistance.

Literature Cited

1. *The Wealth of India. A Dictionary of Indian Raw Materials and Industrial Products;* Raw materials; C.S.I.R.: New Delhi, **1952**; Vol.3, pp 168-170.
2. Inohara, H.; Raz, A. *Glycoconj. J.* **1994**, 11, 527-532.
3. Block, G.; Schwarz, R. In *Natural Antioxidants in Human Health and Disease;* Frei, B., Ed.; Academic Press: New York, **1994**; pp 129-155.
4. Lundholm, K.; Gelin, J.; Hyltander, A. *et al. Cancer Research* **1994**, 54, 5602-5606.
5. Sur, P.; Chatterjee, S.P.; Roy, P.; Sur, B. *Cancer Lett.* **1995**, 94, 27-32.
6. Valette, A.; Gas, N.; Jozan, S.; Roubinet, F.; DuPont, M.A.; Baynard, F. *Cancer Res.* **1987**, 47, 1615-1620.
7. Sur, P.; Matsuo, Y.; Otani, T.; Minowada, J. *Tumori* **1993**, 79, 433-438.
8. Collins, S.J.; Bodner, A.; Ting, R.; Gallo, R.C. *Int. J. Cancer* **1980**, 25, 213-218.
9. Olsson, I.; Gullegerg, U.; Lantz, M. *Leukemia Res.* **1990**, 14, 711-713.
10. Craig, R.W.; Frankfurt, O.S.; Sakagami, H.; Takeda, K.; Bloch, A. *Cancer Res.* **1984**, 44, 2421-2429.
11. Burres, N.S.; Cass, C.E. *Cancer Res.* **1987**, 47, 5059-5064.
12. Ohkami, H.; Tazawa, K.; Yamashita, I.; Shimizu, T.; Murai, K.; Kobashi, K.; Fujimaki, M. *Jpn. J. Can. Res.* **1995**, 86, 523-529.
13. Hardman, W.E.; Cameron, I.L. *Carcinogenesis* **1995**, 16, 1425-1431.
14. Goodwin, J.S.; Ceuppens, J.L. In *Prostaglandin Cellular Immunity and Cancer;* Goodwin, J.S., Ed.; **1985**; pp 1-34.

Chapter 11

Beneficial Effect of Moderate Alcohol Consumption to Health: A Review of Recent Results on Wines

Michikatsu Sato

Wines and Spirits Research Center, Mercian Corporation, 9-1, Johnan 4-chome, Fujisawa 251, Japan

Numerous reports have shown that moderate alcohol consumption has an apparent protective effect on the incidence of mortality from coronary heart disease (CHD). Several large population studies have shown a U shaped relation between alcohol intake and mortality for both men and women. Moderate alcohol intake (10~30 g/day as pure alcohol) reduces cardiovascular mortality by 20%~80% (mean about 50%). There is no question that heavy drinking and total abstinence are not good for health, while moderate alcohol drinking is obviously good for health. Reports on the red wine's beneficial effect on health are increasing. The benefits of red wine ingestion may be due in part to the alcohol, but largely to the presence of phenolics in abundance. There is so much data on the prevention of oxidation of low-density lipoproteins (LDL) by wine phenolics, and on the inhibition of platelet aggregation, reducing the risk for thrombosis. We showed a direct correlation between the phenolic content, and the ability of the wine constituents to scavenge superoxide radicals. In this review, the benefits of wine ingestion, including our recent results on its superoxide radical scavenging activities, are described. The constituents contributing most to health are also discussed from recent studies reported.

The wine consumption in Japan is rapidly increasing in these 3 years, and the annual increase rate is around 10 to 20%. Of wines, the red wine market is especially developing sharply. The reason for the increase in the wine market in Japan may be largely due to the appearance of reasonable priced wines, with acceptable quality. Connection of the healthy image of red wine ingestion to the recent movement seeking a healthy diet, seems to result in the increased ratio of red wine. The trend is also developing in East Asian countries. Wine consumption in Thailand, China, Taiwan, etc., is increasing. The concept of prevention of diseases related to life style, by having proper diet has been gradually accepted. Wine seems to be integrated into the Japanese diet as a part of proper foods.

Here, the benefits of moderate consumption of alcohol, antioxidant effects of wine phenolics to low density lipoprotein (LDL), and the superoxide radical

114

scavenging activities of wine are described including our studies and recent scientific papers on the benefits of wine ingestion.

Effects of Moderate Alcohol Consumption

Numerous data (*1-6*) have shown that moderate alcohol consumption has an apparent protective effect on the incidence, or mortality, from coronary heart disease (CHD). Several large population studies have shown a U shaped relation between alcohol intake, and mortality, for both men and women. Moderate alcohol intake (10-30 g/day as pure alcohol) reduces cardiovascular mortality by 20%~80% (mean about 50%) (*7*). There is no question that heavy drinking and total abstinence are not good for health, while moderate alcohol drinking is obviously good for health.

A moderate intake of alcohol results in a reduction of platelet aggregation (*8-13*). Coronary thrombosis can be prevented by the inhibition of platelet aggregation activity. Alcohol consumption in moderation decreases the LDL, and increases HDL level in blood (*12-14*). The LDL reduction effect has a great advantage in reducing the risk of atherosclerosis, because a high level of plasma LDL tends to be oxidized due to the long duration in blood. Oxidation of LDL is the signal of incorporation by monocytes/macrophages (*15, 16*). The LDL oxidation is the initiation step of arteriosclerosis. Alcohol reduces the release of catecholamines in response to stresses (*17*). As stresses generate the harmful superoxide radicals in human body, it is important for the individual to have a stress release, to reduce the risks of various diseases. Heavy drinking is obviously hazardous to the human body, but proper or moderate drinking is good for health.

'French Paradox' and Effects of Phenolics in Red Wine

Reports on the red wine's beneficial effects on health are increasing (*1, 2, 18, 19*). The benefits of red wine ingestion may be due, in part, to the alcohol, but largely due to the presence of the abundance of phenolics. A comparative study, initiated by the WHO, has shown a marked difference in mortality and morbidity from CHD among, especially, French and US populations (*20*). Despite a similar intake of a saturated fatty acid diet, and comparable levels of plasma cholesterol content, the French subjects were less susceptible to CHD than the US subjects. Renaud and de Lorgeril (*8*) postulated that consumption of wine was the only dietary factor responsible for this discrepancy, commonly referred to as the 'French Paradox.' There has been so much data on the prevention of oxidation of low-density lipoprotein (LDL) by wine phenolics (*19, 21-23*) and on the inhibition of platelet aggregation, reducing the risk for thrombosis (*9, 24-26*).

The effect of red wine ingestion on the antioxidant activity of serum, and on the inhibition of LDL oxidation has been examined using human volunteers. Maxwell *et al.* (*27*) administered Bordeaux red wine (5.7 mL/kg) with meals, to 10 healthy students, and the antioxidant activity of serum was periodically measured. The serum antioxidant activity rose rapidly after ingestion of red wine, and the activity reached a peak after 90 min. Although the antioxidant activity gradually declined after the peak, the antioxidant levels were still significantly raised at 4 hours. The report may be the first study using human volunteers to examine the effect of red wine ingestion on the antioxidant status of human blood. In Japan, Kondo *et al.* (*22*) reported the inhibition effect of red wine ingestion on human LDL oxidation, using 10 healthy volunteers. Bordeaux red wine (0.8 g/kg ethanol per day) was administered for two weeks, and compared the lag time to oxidize the plasma LDL. The lag time of the red wine ingested group was significantly longer than that of non-drinking group, and that of before drinking wine. The result directly demonstrated that regular, and long-term consumption of red wine, but not ethanol, inhibited LDL oxidation *in vivo*. Whitehead *et al.* (*28*) also examined the serum antioxidant status after ingestion of 300 mL of red or white wine using 18 volunteers. The serum antioxidant activity significantly increased after one hour (18% increase) and two hours (11% increase) in the group (n = 9) of red wine ingestion.

Large numbers of ecological data and *in vivo* administration data show that red wine ingestion has beneficial effects on health, not only for the reduction of risk of atherosclerosis and CHD, but also reduction of overall mortality.

Which Component of Red Wine Is Really Contribute to the Reduction of Cardiovascular Disease?

There is a debate on which beverage offers the most protection from cardiovascular diseases (CVD) among alcoholic products. Recently, Rim reviewed papers on ingestion of wines, beer, and spirits in his invited commentary (*29*). The results of the 12 case-control and cohort studies were reviewed, and the same number of studies (seven) found an inverse association between alcohol and CVD for each type of alcoholic beverages. Three of the studies reported that only wine consumption was inversely associated with CVD. However, wine was exclusively consumed in the 2 studies out of 3 cases. He concluded that "From all of the observational data collected, from more than 305,000 men and women, the available evidence suggests that no particular type of beverage provides substantial additional cardiovascular benefit apart from its ethanol content."

Limiting to the preventive effect against CVD, alcohol itself may be the most important factor in alcohol beverages. However, red wine contains a lot of phenolics, and phenolics efficiently scavenge superoxide radicals (*30*). Now various diseases have been found to be caused by the action of reactive oxygen species such as superoxide radicals (*31-33*). Therefore, I believe that red wine ingestion in moderation should have additional benefits for health.

Superoxide Radical Scavenging Activities of Wines

We thought that the antioxidant effect of free radicals may be the most important factor for the beneficial effects of wine. Generally speaking, ischemia and the following reperfusion cause the formation of superoxide radicals. Stress, smoking, drug and food intake also generate free radicals. In the human body, there are defense systems to protect themselves from the attack of free radicals. Superoxide dismutase (SOD), glutathione peroxidase, vitamin C, E, carotenoids, etc., in our body can quench or scavenge free radicals. The excess amount of free radicals oxidizes lipoprotein, lipids, protein, DNA (*34*), etc., and the damages caused thereby lead to various diseases such as CHD, arteriosclerosis, cancer, aging, etc. (*31*). Hence, we (*30*) have investigated the relationship between the phenolic content of wine, and the superoxide radical scavenging activity (SOSA).

Forty-three wine samples, differing in their vintage, variety, and region of production, were collected, and the relationship between superoxide radical scavenging activity (SOSA), and constituents in wines was examined (*35*). There was no correlation between the SOSA values and the free SO_2 levels in wine, and also no correlation between the SOSA values and the total SO_2 levels in wine. There was a positive correlation between SOSA values and the wine color intensity (absorbance at 520 nm), and the correlation coefficient, $r = 0.7517$. The relationship between SOSA and the total phenol contents in wine was very high ($r = 0.9686$) by a single regression analysis.

The wine produced by Markham Vineyards in Napa Valley, California, showed the most potent SOSA of 1189.1 unit/mL. The next is Barolo produced in Italy. The wines made from Cabernet Sauvignon or Nebbiolo had a high phenolic content and SOSA. The wines from Tempranillo, Pinot noir, and Merlot were the next group. The wines from Gamay showed low SOSA values in red wines. White wines had a low phenolic content and SOSA values. Japanese domestic wines made from Cabernet Sauvignon and Merlot also had high SOSA values.

We further examined which of the wine components contributed most to superoxide radical scavenging activity (36). Twelve varieties of wines were chosen as representatives, and were fractionated with a Sep-pak cartridge. The relationship between the total polyphenol contents and the SOSA values in the 12 wines was very close, and the correlation coefficient was 0.97. There was almost no correlation (r = 0.27) between the polyphenol contents and the SOSA values in the fraction (A) containing phenolic acids, sugars, organic acids, amino acids and salts. The correlation between the polyphenol contents and the SOSA values in the fraction (B) containing catechin, procyanidins and flavonols, was very close, and the r value was 0.957. The relationship in the fraction (C) containing anthocyanin monomers, polymers and tannins, was very close and the correlation coefficient was 0.978. Of the fractions, the fraction B exhibited about half of the polyphenol content and SOSA compared with the fraction C. Therefore, the fraction C containing anthocyanins and tannins is believed to contribute most to the SOSA value. We further examined the relation of SOSA values and components in each fraction by HPLC analysis. From the analysis, anthocyanin polymers, but not monomers contained in fraction C had the highest correlation coefficient with SOSA values.

As anthocyanin polymers might contribute most to SOSA, we examined the formation of oligomers between anthocyanin and catechin in a model wine solution (37). As the grape berries of *Vitis vinifera* most abundantly contain malvidin-3-glucoside, we isolated the anthocyanin monomer from dried grape skins. The formation of anthocyanin oligomers between the malvidin-3-glucoside and (-)-epicatechin, in the presence of acetaldehyde, was examined in a model wine solution. Novel peaks appeared by 4-day incubation, and the peaks were clearly detected at 9 days' incubation. The polymerization was also occurred in the absence of acetaldehyde, but the formation rate was much slower. Compounds corresponding the two peaks were isolated by HPLC, and the inhibition activity against human platelet aggregation was examined. Both of the compounds showed three to four times higher anti-platelet aggregation activity than malvidin-3-glucoside. Epicatechin had not the activity. Superoxide radical scavenging activity (SOSA) of the compounds was examined in the hypoxanthin-xanthin oxidase system. The SOSA of the compounds was about 4 times higher than malvidin-3-glucoside, and the activity was comparable with epicatechin. The peaks were also clearly detected from commercially available red wine. The detailed results will be published elsewhere in the near future. Now, we are postulating that the most contributing factor to its antioxidant activity in red wine will be oligomers of anthocyanins, and the polymerization may be occurred between anthocyanins and catechins.

Recent Studies on Benefits of Red Wine Ingestion

Red wine contains another beneficial substance for health, named resveratrol. Resveratrol, 3,5,4'-trihydroxystilbene, is one of phytoalexin compounds produced by grapevines in response to fungal infection, or abiotic stresses, such as UV irradiation (38, 39). *trans*-Resveratrol is attracting attention, because it is one of the components present in wines that could be responsible for the decrease in CHD observed among wine drinkers. The effect could be due to its inhibition ability of

LDL oxidation (*19*), the anti-platelet aggregation activity (*40, 41*), and the inhibition of eicosanoid synthesis (*41, 42*). Wines contain piceid, 3-b-glucoside of resveratrol, and it also blocks platelet aggregation (*42, 43*). The piceid will release resveratrol by b-glucosidase digestion in the intestine (*44*). The *cis*-isomers have anti-tumor activity as well as the trans isomers by inhibiting protein-tyrosine kinase (*45*).

Very recently, the potent anti-cancer effect of *trans*-resveratrol was reported by a group of the University of Illinois (*46*). In the report, resveratrol had anti-initiation, anti-promotion, and anti-progression activities against cancer cells *in vitro*. It also inhibited the development of preneoplastic lesions in carcinogen-treated mouse mammary glands in culture and inhibited tumorigenesis in a mouse skin cancer model.

The resveratrol contents in wines of various grape origins and varieties have been reported (*47-50*). The resveratrol level in wines is not high, e.g., 1-10 mg/L in red wines, and ‹1 mg/L in white wines. We also investigated resveratrol and piceid levels, along with their isomers, in wines produced from domestically grown grapes (Sato, M., Suzuki, Y. Okuda, T., Yokotsuka, K. *Biosci. Biotech. Biochem.* in press). The highest total stilbene content including *trans*-, *cis*-resveratrol, *trans*-, and *cis*-piceid was 13.4 mg/L in Pinot noir wine. As the effective dose of resveratrol reported ranging 3.1-27 µM (*46*), the levels contained in red wine seem to be too low to be effective. However, regular ingestion of red wine may have some effects due to the resveratrol.

Red wine inhibits the growth of *Helicobacter pylori in vitro*. Fuglsang and Muller (*51*) reported that California Burgundy inhibited the *H. pylori* growth within 15 min. It may be related to stomach cancer prevention by daily red wine ingestion.

Very recently, Orgogozo *et al.* (*52*) reported that the odds ratio was 0.18 (p ‹ 0.01) for the incidence of dementia and 0.25 (p ‹ 0.03) for Alzheimer's disease in the elderly subjects aged 65 and over, drinking 3 to 4 standard glasses of red wine (›250 and up to 500 mL). He described that as the subjects living in the district of Gironde and Dordogne were drinking red wine exclusively, the effect by red wine ingestion is not conclusive. However, there may be some benefits for elderly people over 65 years old. As the reactive oxygen species play an important role for neurodegenerative disorders include Alzheimer's disease, etc. (*53*). Phenolics in red wine might be effective against the diseases.

Concluding Remarks

Moderate alcohol consumption (10-30 g/day as pure alcohol) is obviously beneficial for health, and the reduction of risk against cardiovascular diseases (CVD) will be due to the ethanol itself contained in beverages. Red wine contains a lot of phenolics, and it efficiently scavenges superoxide radicals, with phenolics contained in red wine. Red wine ingestion elevates plasma antioxidant activity *in vivo*. Of the components in red wines, anthocyanin oligomers seems to contribute most to the superoxide radical scavenging activity (SOSA). Polymerization between anthocyanin and epicatechin was demonstrated in a model wine, and the oligomers were detected in commercially available wines.

Regular and proper amount ingestion of red wine might be beneficial against diseases such as dementia, Alzheimer's diseases, stomach cancer, etc. other than CVD. Although I do not recommend drink wine to abstainers, if you drink alcohol, please drink red wine in moderation with a meal enjoying your life.

Literature Cited

1. Sr. Leger, A.S.; Cochrane, A.L. *Lancet* I. **1979**, 1017-1020.
2. Friedman, L.A.; Kimball, A.W. *Am. J. Epidemiol.* **1986**, *124*, 481-489.
3. Marmot, M.G.; Rose, G.; Shipley, M.J. *Lancet* I, **1981**, 580-583.
4. Kono, S.; Ikeda, M.; Tokudome, S.; Ogata, M. *Int. J. Epidemiol.* **1986**, *15*, 527-532.
5. Cullen, K.J.; Knuiman, M.W.; Ward, N.J. *Am. J. Epidemiol.* **1993**, *137*, 242-248.
6. Sharper, A.G.; Wannamethee, G.; Walker, M. *Lancet* ii. **1988**, 1267-73.
7. Griffith, M.J. *Br. Heart J.* **1995**, *73*, 8-9.
8. Frankel, E.N.; Kanner, J.; German, J.B.; Parks, E.; Kinsella, J.E. *Lancet* **1993**, *341*, 454-57.
9. Lee, A.J.; Smith, W.C.; Lowe, G.D.; *et al. J. Clin. Epidemiol.* **1990**, *43*, 913-919.
10. Renaud, S.C.; Ruf, J.-C. *Clinica. Chimica. Acta* **1996**, *246*, 77-89.
11. Seigneur, M.; Bonnet, J.; Dorian, B.; Benchimol, D.; Drouillet, F.; Gouverneur, G.; Larrue, J.; Crockett, R.; Boisseau, M.; Ribereau-Gayon, P.; Bricaud, H. *J. Appl. Card.* **1990**, *5*, 215-22.
12. Klatsky, A.L. *Alcohol. Hlth. Res. Wld.* **1990**, *14*, 289-300.
13. Criqui, M.H. *Br. J. Adduct.* **1990**, *85*, 854-8.
14. Rankin, J.G. *Contemp. Drug Prob.* **1990**, *21*, 45-57.
15. Brown, M.S.; Goldstein, J. L. *Ann. Rev. Biochem.* **1983**, *52*, 223-261.
16. Brown, M.S.; Goldstein, J. L. *Science* **1986**, *232*, 34.
17. Pohorecky, L.A. *Alcohol* **1986**, *7*, 537-46.
18. Renaud, S.C.; de Logeril, M. *Lancet* **1992**, *339*, 1523-1526.
19. Frankel, E.N.; Kanner, J.; German, J.B.; Parks, E.; Kinsella, J.E. *Lancet* **1993**, *341*, 454-457.
20. NRC, In *Diet and Health*; National Research Council, National Academy Press: Washington, D.C. , 1989.
21. Frankel, E.N.; Waterhouse, A.L.; Kinsella, J.E. *Lancet* **1993**, *341*, 1103-1104.
22. Kondo, K.; Matsumoto, A.; Kurata, H.; Tanahashi, H.; Koda, H.; Amachi, T.; Itakura, H. *Lancet* **1994**, *344*, 1152.
23. Fuhrman, B.; Lavy, A.; Aviram, M. *Am. J. Clin. Nutr.* **1995**, *61*, 549-554.
24. Renaud, S.C.; Beswick, A.D.; Fehily, A.M.; Sharp, D.S.; Elwood, P.C. *Am. J. Clin. Nutr.* **1992**, *55*, 1012-1017.
25. Seigneur, M.; Bonnet, J.; Dorian, B.; Benchimol, D.; Drouillet, F.; Gouverneur, G; Larrue, J.; Crockett, R.; Boisseau, M.; Ribereau-Gayon, P.; Bricaud, H. *J. Appl. Card.* **1990**, *5*, 215-222.
26. Ruf, J.-C.; Berger, J.-L.; Renaud, S. *Arterioscler. Thromb. Vasc. Biol.* **1995**, 140-144.
27. Maxwell, S.R.J.; Cruickshank, A.; Thorpe, G. *Lancet* **1994**, *344* (July), 193-4.
28. Whitehead, T.P.; Robinson, D.; Allaway, S.; Syms, J.; Hale, A. *Clin. Chem.* **1995**, *41/1*, 32-35.
29. Rim, E.B. *Am. J. Epidemiol.* **1996**, *11*, 1094-1098.
30. Sato, M.; Ramarathnam, N.; Suzuki, Y.; Ohkubo, T.; Takeuchi, M.; Ochi, H. *J. Agric. Food Chem.* **1996**, *44*, 37-41.
31. Marnett, L.J.; Hurd, H.K.; Hollstein, M.C.; Levin, D.E.; Esterbauer, H.; Ames, B.N.: *Mutat. Res.* **1985**, *148*, 25-34.
32. Cutler, R.G. In *Free Radicals in Biology*; Pryor, W.A. Ed.; Acad. Press: New York, **1984**, *6*, 371-428.
33. Aruoma, O.I.; Halliwell, B. In *Free Radicals and Food Additives*; Taylor and Frances: London, 1991.

34. Kehrer, J.P. Free radicals as mediators of tissue injury and disease. *CRC Crit. Rev. Toxicol.* **1993**, *23*, 21-48.
35. Sato, M.; Ramarathnam, N.; Suzuki, Y.; Ohkubo, T.; Takeuchi, M.; Ochi, H. *ASEV Jpn. Rep.* **1995**, *6*, 233-236.
36. Sato, M.; Ramarathnam, N.; Suzuki, Y.; Ohkubo, T.; Takeuchi, M.; Ochi, H. In *Abstract papers of International Conference on Food Factors* (Hamamatsu), 1995, p. 81.
37. Morimitsu, K.; Koike, S.; Suzuki, Y.; Sato, M.; Osawa, T. Abstract Papers, Annual Mtg. of Agric. Chem. Soc. Japan, 1997. p. 58.
38. Langcake, P.; Pryce, R.J. *Physiol. Plant Pathol.* **1976**, *9*, 77-86.
39. Langcake, P.; Pryce, R.J. *Phytochemistry* **1977**, *16*, 1193-1196.
40. Bertelli, A.A.E.; Giovannini, L.; Giannesi, D.; Migliori, M.; Bernini, W.; Fregoni, M.; Bertelli, A. *Int. J. Tissue React.* **1995**, *17*, 1-3.
41. Pace-Asciak, C.R.; Hahn, S.E.; Diamandis, E.P.; Soleas, G.; Goldberg, D.M. *Clin. Chim. Acta* **1995**, *235*, 207-219.
42. Kimura, Y.; Okuda, H.; Arichi, S. *Biochim. Biophys. Acta* **1985**, *834*, 275-278.
43. Shan, C.W.; Yang, S.Q.; He, H.D.; Shao, S.L.; Zhang, P.K. *Acta Pharmacol. Sin.* **1990**, *11*, 527-530.
44. Hackett, A.M. In *Plant Flavonoids in Biology and Medicine: Biochemical Pharmacological and Structure-Activity Relationships; Progress in Clinical and Biological Research 213*; Cody V.; Middleton, E.J.; Harborne, J.B.; Eds., Liss, New York, 1986, pp. 177-194.
45. Jayatilake, G.S.; Jayasuriya, H.; Lee, E.S.; Koonchanok, N.M.; Geahlen, R.L.; Ashendel, C.L.; McLaughlin, J.L.; Chang, C.J. *J. Nat. Prod.* **1993**, *56*, 1805-1810.
46. Jang, M.; Cai, L.; Udeani, G.O.; Slowing, K.V.; Thomas, C.F.; Beecher, C.W.W.; Fong, H.H.S.; Farnsworth, N.R.; Konghorn, A.D.; Mehta, R.G.; Moon, R.C.; Pezzuto, J.M. *Science* **1997**, *275*, 218-220.
47. Jeandet, P.; Bessis, R.; Maume, B.F.; Sbaghi, M. *J. Wine Res.* **1993**, *4*, 79-85.
48. Lamuela-Raventós, R.M.; Romero-Pérez, A.; Waterhouse, A.L.; de la Torre-Boronat, M.C. *J. Agric. Food Chem.* **1995**, *43*, 281-283.
49. Lamuela-Raventós, R.M.; Waterhouse, A.L. *J. Agric. Food Chem.* **1993**, *41*, 521-523.
50. Vrhovsek, U.; Eder, R.; Wendelin, S. *Acta Alimantaria* **1995**, *24*, 203-212.
51. Fugelsang, K.C.; Muller, C.J. In *Proc. Symp. on Wine and health*; Waterhouse, A.L.; Rantz, J.M., Eds.; Am. Soc. Enol. Vitic., Davis, CA, 1996, pp. 43-45.
52. Orgogozo, J.-M.; Dartigues, J.-F.; Lafont, S.; Letenneur, L.; Commenges, D.; Salamon, R.; Renaud, S.; Breteler, M.B. *Rev. Neurol.* (Paris), **1997**, *153* (3), 183-192.
53. Knight, J.A. *Ann. Clin. Lab. Sci.* **1997**, *27*, 11-25.

Vegetables and Related Compounds

Chapter 12

Anti-Fatigue Effects of Yiegin

Shiyun Yan[1], Jiyan Zhou[1], and Hirotomo Ochi[2]

[1]Shanghai Chinese Medicine University, 530 Lingling Road, Shanghai 200032, Peoples Republic of China
[2]Japan Institute for the Control of Aging, 723-01 Haruoka, Fukuroi City, Shizuoka 437-01, Japan

Fatigue is caused by excessive physical or mental stress or due to drug intake, environmental insult, nutritional imbalance and diseases. In this paper, several different experiments were designed to evaluate the anti-fatigue mechanism and clinical functions of yiegin. We found that yiegin can enhance exercise capacity; prolong the survival time in oxygen deficiency conditions, increase anti-cold capacity; and activate the sperm generation system.

Fatigue is caused by excessive exercise, which could be physical or mental, or due to drug intake, environmental insult, nutritional imbalance and diseases(1,2). Fatigue makes it difficult for the body to continue the normal metabolism or maintain exercise strength. It also results in a decrease in the activity of the body's defense system. Our study is based on the traditional Chinese-medical theory, which focuses on balancing the effects of Chinese-medicine by elevating the body function capacity, and stimulating body energy metabolism(3). Therefore, we carried out several different experiments to evaluate the anti-fatigue mechanism and clinical efficacy of yiegin, a brownish-black, high density liquefied Chinese drug.

Materials and Methods:

Materials: Kunming mice (body weight from 18 to 22 g, with equal numbers of male and female, China) were provided by the Experimental Animal Center of the Shanghai Medical University. Other chemicals were of the highest purity commercially available. Yiegin was provided by Shanghai Chinese Medicine University, and used at concentration of 1 g crude drug/mL. Another Chinese drug called Zhuangyaojianshenwan (Guangzhou Chen Li Ji Pharmaceutical Company) was used in this study as an internal reference. The crude drugs were suspended in water and orally administered to mice at 8 g/kg body weight basis. Water was used as placebo.

Physical fitness test. Mice were divided to three groups, namely the placebo group (water), Yiegin group (8 g crude drug/kg body weight basis) and Zhuangyaojianshenwan group (8 g crude drug/kg body weight basis). The supplements were orally administered once a day, and continued for 10 days. After 1 hour from the last feed, lead weights, constituting 5% of the body weight, was tagged to the tail, and the mice were forced to swim in a 25, 30 cm deep water bath (4). The total swimming time of each group was determined.

Oxygen deficiency test. Similar to the experimental design described above, mice were divided into three groups. The supplements were orally administered once a day and continued for 8 days. Then after 1 hour from the last feed, each mouse was placed in a 250 mL sealed bottle containing 15 g oxygen absorbent. The total survival time of each mouse was determined.

Cold proof test. Mice were divided into three groups. The supplements were orally administered once a day and continued for 8 days. Then after 1 hour from the last feed, mice were placed in a room at -4 for 4 or 6 hours. The survival rate of each group was determined.

Changes of blood urea nitrogen level. Mice (body weight 18-22 g) were divided to three groups. The supplements were orally administered once a day and continued for 10 days. Then after 30 minutes from the last feed, mice were subjected to swimming exercise in a 30 water bath for 90 minutes, and blood was withdrawn for urea nitrogen test.

Enhancement of sperm generation capacity. Twelve months old male mice (body weight 35-41 g) were divided into three groups. The supplements were orally administered once a day, and continued for 10 days. Testes were removed, weighted and homogenized at 5 mg/mL with buffered culture medium. The quantity and physiological activity of the sperm were determined under microscope.

Results and Discussion:

First, we examined the effect of yiegin on the enhancement exercise capacity by a physical fitness test. Compared to the placebo group (water), the average swimming time of yiegin treated groups and commercial Chinese drug treated groups were 397.378.6 and 363.352.6 minutes (Table I). Increases in enhancement exercise capacity for both groups were 50.0% and 35.9% respectively. These results indicate that yiegin has a significant effect on the enhancement of exercise capacity.

124

Table I. Effect of yiegin on the enhancement of exercise capacity.

Treatments	Average swimming time (min)	Extension percent
Yiegin(n=10)	397.3 8.6 (b)	50.0
*CSCD (n=10)	363.3 2.6 (a)	35.9
Placebo(Water) (n=10)	267.1 92.8	

*CSCD: Commercially Sold Chinese Drug

(a) $p<0.05$; (b) $p<0.01$

We then tested the effect of yiegin on prolonging the survival time of mice, under an oxygen deficiency state. As shown in Table II, the average survival time for yiegin, commercially sold Chinese drug (Zhuangyaojianshenwan), and the placebo (water) groups were 48.66.4, 52.58.2, and 32.27.2 minutes respectively. Compared to the placebo (water) group, the extension of yiegin group was 49.6%, although the CSCM group showed a higher survival time, 63.3%. This result also indicated that yiegin prolongs the survival time of mouse under an oxygen deficiency state.

The results obtained in a cold proof test, to examine the effect of yiegin on the enhancement of anti-cold capacity, is shown in Table III. The survival percentage of yiegin group, commercially sold Chinese drug

Table II. Effect of yiegin on prolonging the survival time of mouse in an oxygen deficiency state.

Treatments	Average survival time (min)	Extension percent
Yiegin(n=10)	48.6 6.4	49.6
*CSCD (n=10)	52.5 8.2	63.3
Placebo(Water) (n=10)	32.2 7.2	

*CSCD: Commercially sold Chinese Drug

(Zhuangyaojianshenwan) group and placebo (water) group, following a 4 hour test period, were 60, 20 and 10% respectively. For a 6 hour test period, the survival percentage of yiegin, commercially sold Chinese drug (Zhuangyaojianshenwan) and placebo (water) group were 70, 50 and 10% respectively. The results of both drug treated groups showed that yiegin significantly enhanced the anti-cold capacity in mice.

Table III. Effect of yiegin on the enhancement of anti-cold capacity.

Treatments	Survival percent	p
4 hr. test time		
Yiegin(n=10)	60	<0.05
*CSCD (n=10)	20	
Placebo(Water) (n=10)	10	
6 hr. test time		
Yiegin(n=10)	70	<0.01
*CSCD (n=10)	50	<0.05
Placebo(Water) (n=10)	10	

*CSCD: Commercially sold Chinese Drug

Table IV shows the results of the changes of blood urea nitrogen before, and after drug supplementation. Compared with the blood urea nitrogen level before supplementation, the placebo group did not show any change, but the yiegin treated group showed a decrease from 6.58 to 3.32(almost a 50% decrease). The commercially sold Chinese medicine (Zhuangyaojianshenwan) treated group also showed a decrease from 6.41 to 4.97. This result indicates that yiegin can significantly stimulate metabolism.

Table IV. Effect of yiegin on the variation in blood urea nitrogen level.

Treatments	BUN** Before supplement		BUN** After supplement		p
Yiegin	6.58	1.34	3.32	2.74	<0.05
*CSCD	6.41	1.55	4.97	2.60	<0.1
Placebo(Water)	6.53	1.43	6.54	1.30	>0.05

*CSCD:Commercially Sold Chinese Drug
**BUN: Blood Urea Nitrogen

The enhancement effect of yiegin on sperm generation capacity is shown in Table V. Compared to the placebo (water) group, 4 g/kg yiegin administrated group showed an increase in sperm quantity, but not so much in activity. However, 8 g/kg yiegin administered group showed a significant increase both in sperm quantity, from 80 to 122 million/100mg and activity, from 65.5 to 82.5.

Table V. Effect of yiegin on the enhancement of sperm generation capacity.

	Sperm		
Treatments	Quantity(Million/100mg)		Activity
Yiegin 4g/kg	99.0	19.2 (a)	65.5 13.2
Yiegin 8g/kg	122.5	26.7 (b)	82.5 6.4
Placebo(Water)	80.2	21.1	65.5 4.6

(a)$p<0.05$;(b)$p<0.01$

From these results, we conclude yiegin possesses the capacity to (a) enhance exercise capacity; (b) prolong the survival time in oxygen deficiency state; (c) enhance anti-cold capacity; and (d) activate sperm generation system.

Acknowledgments: The authors thank Dr. Rongzhu Cheng and Dr. Sachi Sri Kantha (Japan Institute for the Control of Aging) for critical reading and assistance in the preparation of this manuscript.

Literature cited:
1. Yan Weiyi and Liang Shangrong. *J. of China Sports Medical Science.* 1994, **13 (1)**, 28.
2. Gu Haiqian. *J. of Chinese Medicine.* 1988, **19(5)**, 37.
3. Cheng Shide and Mong Jinchun (eds.) *The Theory of Chinese Medicine,* Shanghai, 1984.
4. Xie Minhao. *J. of China Sports Medical Science,* 1988, **7(1)**, 38.

Chapter 13

A Potent Antioxidative and Anti-UV-B Isoflavonoids Transformed Microbiologically from Soybean Components

Akio Mimura[1], Shin-ichi Yazaki[1], and Hiroshi Tanimura[2]

[1]Department of Biotechnology, Yamanashi University, Takeda, Kofu 400, Japan
[2]Kobe Steel Ltd., Kobe 651, Japan

Japanese traditional fermented soybean foods (miso(soybean paste), soy sause, natto and so on) have been hypothesized to contribute to the lower incidence of human cancers and cardiac diseases. Soybeans are rich in isoflavonoid glucosides such as daidzin and genistin. During the fermentation with microorganisms, these glucosides can be hydrolized to aglycon isoflavones (daidzein and genistein), and further transformed to biologically active compounds such as more hydroxylated isoflavones. Several kinds of fungi relating to the fermented foods and bacteria isolated from soil were screened for the production of potent activity of antioxidation (anti-UV-B) from soybean components. *Aspergillus niger* IFO 4414 was selected as the most potent producer of antioxidative isoflavones. The fungus was cultivated in the medium composed of soybean flour, and it was observed that anti-UV-B activity of the culture extracts was increased remarkablly during the fermentation. From the fermented soybeans, a isoflavone with potent anti-UV-B activity was isolated and identified as 4',7,8-trihydroxyisoflavone (8-hydroxydaidzein), which was demonstrated as the hydroxylated product of daidzein at the 8-position of A-ring. The maximum conversion rate to 4',7,8-trihydroxyisoflavone from daidzein was 67.8%(w/w). 4',7,8-trihydroxyisoflavone was observed to have almost same anti-UV-B activity (antioxidative activity) as BHA, 60 to 100 times stronger activity than alpha-tocopherol, and about 15 times stronger activity than daidzein and genistein, using the measurement method with rabbit erythrocyte membrane ghosts irradiating UV-B light.

127

Soybeans are known to contain several isoflavones and their glucosides (1,2). These compounds have been demonstrated to have antioxidative activity (3,4). The antioxidative activity of soybeans is considered to be the biological activity to maintain our good health. The extent of the antioxidative activity of isoflavones is positively correlated to the number of hydroxy groups in the isoflavone molecules (5). From this point of view, more potent antioxidative and anti-UV-B isoflavones can be transformed microbiologically from natural isoflavones in soybeans. A higher consumption of soybean foods, rich in these natural and transformed isoflavones, has been considered to be a factor in a lower incidence of human cancer, cardiac diseases and other age-related disorders.

The Japanese have taken much of the fermented soybean foods, such as miso (soybean paste), soy souse, natto and so on, and observed that during fermentation of soybeans to produce miso and soy sauce, the antioxidatiove activity increased remarkablly. Soybeans contain many kinds of isoflavone glucosides such as daidzin, genistin and glycitin, and during fermentation with microorganisms, these isoflavone glucosides can be transformed to their corresponding aglycone, and further metabolized. An Indonesian soybean fermented food, Tempeh has reported to have more antioxidative activity than its raw materials (6,7), and as a metabolized isoflavone with potent antioxidative activity, 4',6,7-trihydroxyisoflavone has been isolated and identified as the transformed compound from the soybean isoflavones (8,9,10).

The biological activity of soybean isoflavones has been investigated intensively for antioxidative activity, fungistatic action, antihemolytic activity, anti-tumor promoting activity, and anticancer activity (5,11,12,13,14). Higher antioxidative activities in Japanese soybean fermented foods suggest that they contain several kinds of compounds with potent antioxidative activity. In our research with fungi for fermentation of soybeans, a fungus culture which can produce a potent anti-UV-B and antioxidative activities was found. From the culture broth, a microbiologically transformed isoflavone from daidzin in soybeans was isolated and identified its chemical structure.

Screening of fungi transformed soybean components

One handred strains of fungi, including type cultures and newly isolated fungi from soil, were cultivated aerobically in the soybean medium composed of 10% soybean flour (Sigma type 1, pH 6.0) for 4 days at 28 C and the culture brothes were extracted with 80% aceton, and then the extracts were assayed for their anti-UV-B activity. The assay procedure has been reported and shown breifly in Figure 1 (15). The increase of antioxidative activity in typical cultures of fungi were presented in Figure 2. Among them one fungus, *Aspergillus niger* IFO 4414 produced more antioxidative activity than soybean medium without fermentation (indicated as S.B. medium). *Aspergillus oryzae* IFO 4176, *Aspergillus soyae* IFO 4200, *Penicillium carmemberti* IFO 8187, and one isolated fungus 33M-2 indicated the production of anti-UV-B active compounds from soybean components. From these results, *Aspergillus niger* IFO 4414 was used for further experiments.

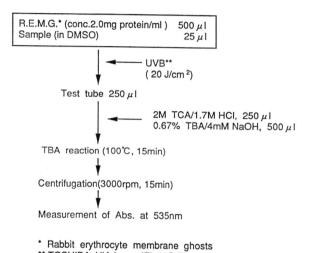

Figure 1. Assay method of anti-UV-B activity.

130

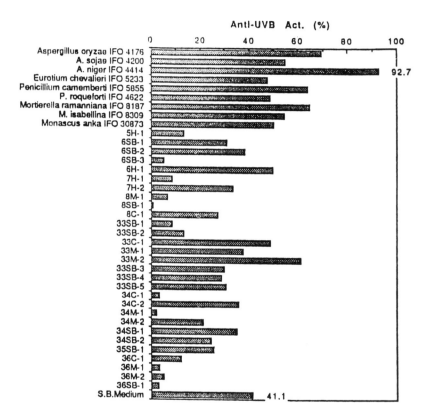

Figure 2. Anti-UV-B activity of various extracts offermented soybeans with fungi.

Purification of Anti-UV-B compounds from *Aspergillus niger* IFO 4414 culture

340 grams of fermented material was extracted with 80% acetone. After the acetone was evaporated under reduced pressure, the aliquot was extracted with ethylether. The ether extract was condensed, and applied to silica-gel chromatography (Silica Gel 60, E.Merck), and eluted with chloroform/methanol. The active fraction (chloroform/methanol, 9:1) was applied to the HW-40F (Toyopearl TSK gel) column chromatography, and eluted with methanol/water. The 40% methanol fraction was applied to further purification with HPLC (Deverosil ODS-5 column, 40% methanol/10mM phosphate buffer pH 2.7, UV 260nm detection). The purified compound, of 50 mg with pale yellow color, was obtained.

The elusidation of chemical structure was performed using UV, IR, MS, and NMR. Analytical results were summarized in Figure 3. From the structure analyses, the active compound was identifed as 4',7,8-trihydroxyisoflavone (8-hydroxydaidzein).

Many kinds of isoflavones have been reported to be produced by cultivation of *Streptomyces* microorganisms on the medium of soybean meal. This was usually used for antibiotic productions with *Streptomyces* (12,13,14,16). Ohmura et al found 4',7,8-trihydroxyisoflavone from fermentation broth of *Streptomyces* sp., isolated from soil. It was considered that some of isoflavones in soybeans would be transformed to 4',7,8-trihydroxyisoflavone during fermentation. However, they have not demonstrated the scheme of transformation reactions from daidzin to 4',7,8-trihydroxyisoflavone.

Anti-UV-B activity of 4',7,8-trihydroxyisoflavone

The anti-UV-B activity of several antioxidative compounds and isoflavones in soybeans was summarized in Figure 4 and Figure 5. Potent anti-UV-B activity, as strong as butylhydroxyanisol (BHA), was observed in 4',7,8-trihydroxyisoflavone. It also showed about 60 to 100 times stronger activity than alpha-tocopherol, and about 15 times stronger activity than daidzein and genistein. The antioxidative activity of these compounds evaluated with the linolenic acid oxidation method showed good correlation to the anti-UV-B activity (data not shown).

Transformation of 4',7,8-trihydroxyisoflavone from daidzin and daidzein

Aspergillus niger (IFO 4414) was cultivated in 100 ml synthetic medium (Czapek-Dox), containing 100 microgram daidzin or daidzein respectively. The production of 4',7,8-trihydroxyisoflavone was analyzed with HPLC (Develosil ODS-5, 35C, methanol/10mM phosphate buffer, UV 260nm detection). Table 1 indicates the transformation reaction of daidzin to 4',7,8-trihydroxyisoflavone. The final yeild of the product was only 2.4 microgram from 100 micrograms of daidzin. However, from the 100 microgram daidzein the final yeild was increased to 15.9 microgram after 10 days cultivation. In this transformation reaction, the final cumulative conversion rate from daidzein to 4',7,8-trihydroxyisoflavone was 15.9%(w/w).

The time course of microbial transformation of daidzin is shown in Figure 6. The results indicate that daidzin was at first hydrolized to its aglycon, daidzein, and

1.Color : pale yellow, powder

2.UV ; λ_{max} (MeOH) nm : 260 (Band II)

3.IR ; ν_{max}(KBr) cm^{-1} : 3450(hydroxyl), 1615(carbonyl), 1580,
1450(aromatic),1275(=C-O-C-)

4. MS m/z (rel. int.) : 270[M$^+$](100), 152[A$_1$$^+$](50), 118[B$_1$$^+$](7)

5.High resolution MS : Found 270.0526, Calcd for $C_{15}H_{10}O_5$ 270.0528

6.^{1}H-NMR(300.13 MHz, DMSO-d_6)
   6.88(2H, d, J=6.7, H-3', H-5'), 6.95(1H, d, J=8.7, H-6),
   7.39(2H, d, J=6.7, H-2', H-6'), 7.48(1H, d, J=8.7, H-5),
   8.31(1H, s, H-2)

7.^{13}C-NMR(75.47 MHz, DMSO-d_6)
   114.0(C-6), 114.7(C-3', C-5'), 115.5(C-5), 117.3(C-8),
   122.5(C-1'), 122.8(C-3), 129.9(C-2', C-6'), 132.7(C-10),
   146.6(C-9), 149.7(C-7), 152.3(C-2), 156.9(C-4'), 174.9(C-4)

Name:
   4', 7, 8-trihydroxyisoflavone

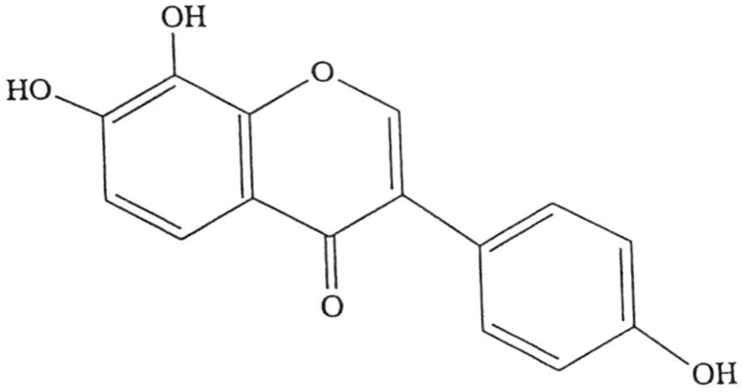

Figure 3. Structure of anti-UV-B isoflavone, 4',7,8-trihydroxyisoflavone (8-hydroxydaidzein).

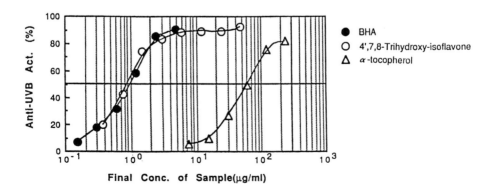

Figure 4. Anti-UV-B activity of 4',7,8-trihydroxyisoflavone.

133

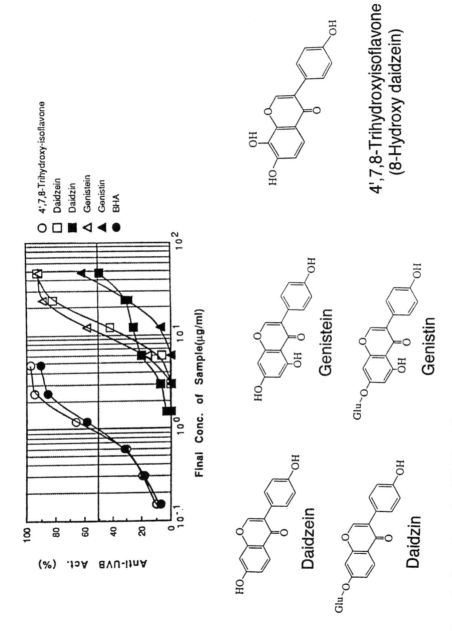

Figure 5. Comparison of anti-UV-B activity among 4',7,8-trihydroxyisoflavone and other isoflavones contained in soybeans.

Table I. Transformation of daidzin to 4',7,8-trihydroxyisoflavone by
Aspergillus niger IFO 441

Cultivation time	Content (µg)	
	daidzin	4',7,8-trihydroxyisoflavone
0 day	100	0
4 days	30.6	0.7
10 days	0.2	2.4

Synthetic medium (Czapek-Dox medium) was used

Table II. Transformation of daidzein to 4',7,8-trihydroxyisoflavone by
Aspergillus niger IFO 4414

Cultivation time	Content (µg)	
	daidzein	4',7,8-trihydroxyisoflavone
0 day	100	0
4 days	80.3	7.7
10 days	62.8	15.9

Synthetic medium (Czapek-Dox medium) was used

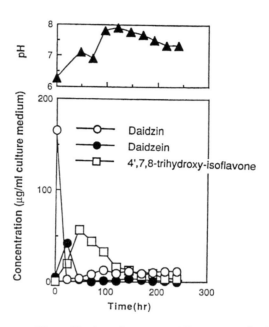

Culture conditions: 20 ml medium / 100 ml Erlen-meyer flask
28°C,rotary shaker (180 rpm.)

Figure 6. Time course of transformation of daidzin to
4',7,8-trihydroxyisoflavone by *Aspergillus niger* IFO 4414.

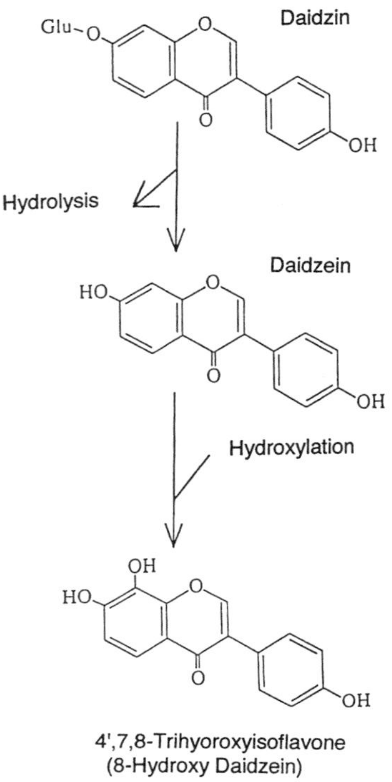

136

Figure 7. Proposed pathway of transformation of isoflavones in soybeans by *Aspergillus niger* IFO 4414.

further oxidized to 4',7,8-trihtydrioxyisoflavon as shown in Figure 7. It was observed that 4',7,8-trihydroxyisoflavone was further metabolized to other unknown compounds during a longer fermentation period (data not shown). In this reaction the maximum conversion rate from daidzin to 4',7,8-trihydroxyisoflavone was 67.8% (w/w) after 48 hour transformation period.

The cell-free extract of *Pseudomonas* sp., grown on (+)-catechin, has been demonstrated to oxidize taxiforin to 3',4',5,7,8-pentahydroxyflavonol (2,3-dihydrogossypetin) as an intermediate in the aerobic metabolism of flavonoids. The enzymatic transformation by taxiforin 8-monooxygenase, in the presence of NAD(P)H and molecular oxygen proceeds the hydroxylation of 8-position of flavonoid A-ring (17).

In this report, it was demonstrated that the hydroxylation of 8-position of the isoflavone A-ring occurs stoichiometrically, by the microbial transformation of *Aspergillus niger* (IFO 4414). From these observations, it can be concluded that isoflavon glucosides in soybeans, such as daidzin, and genistin are transformed microbiologically during fermentation of *Aspergillus niger* (IFO 4414). This type of transformation of isoflavones can occur during the production of soybean fermented foods and more hydroxylated isoflavones could be the origin of the potent antioxidative activity of the soybean fermented foods.

Literature Cited

1. Hammerschmidtt, P. A.; Pratt, D. E. *J. Food Science* **1978,** 43, 556-559
2. Eldridge, A. C. *J. Agric. Food Chem.* **1982,** 30, 353-355
3. Pratt, D. E.; Birac, P. M. *J. Food Science* **1979,** 44, 1720-1722
4. Record, I. R.; Dreosti, I. E.; Mcinerney, J. K. *Nutritional Biochemistry,* **1995,** 6, 481-485
5. Naim, M.; Gestetner, B.; Bondi, A.; Birk, Y. *J. Agric. Food. Chem.* **1976,** 24, 1174-1177
6. Ebata, J.; Fukuda, Y.; Hirai,K.; Murata, K. *Agric. Biol. Chem.* **1972,** 46, 323-329
7. Murakami, H.; Asakawa, T.; Terao, J.; Matsusita, S. *Agric. Biol Chem.* **1984,** 48, 2971-2975
8. Gyorgy, P.; Murata, K.; Ikehata, H. *Nature,* **1964,** 203, 870-872
9. Ikehata, H.; Wakaizumi, M.; Murata, K. *Agric. Biol. Chem.* **1968,** 32, 740-746
10. Hoppe, M.; Chandra Jha, H.; Egge, H. *J. Am. Oil Chem. Soc.* **1997,** 74, 477-479
11. Wei, H.; Wei, L.; Frenkel, K.; Bowen, R.; Barnes, S. *Nutrition & Cancer* **1993,** 20, 1-12
12. Chimura, H.; Sawa, T.; Kumada,Y.; Nagasawa,H.; Matsuzaki, M.; Takita, T.; Hamada, M.; Takeuchi, T.; Umezawa, H. *J. Antibiotics* **1975,** 28, 619-626
13. Komiyama, K.; Funayama, S.; Anraku, Y.; Mita, A.; Takahashi, Y.; Ohmura. S. *J. Antibiotics* **1989,** 42, 1344-1349
14. Funayama, S.; Anraku, Y,; Mita, A.; Komiyama, K.; Ohmura, S. *J. Antibiotics* **1989,** 42, 1350-1355
15. Mimura, A.; Takebayashi, K.; Niwano, M.; Takahara, Y.; Osawa, T.; Tokuda, H. *ACS Symposium Series 547, Food Phytochemicals for Cancer Prevention* II **1994,** 281-294
16. Aoyagi, T. ; Hazato, T. ; Kumagai, M.; Hamada, M.; Takeuchi, T.; Umezawa, H. *J. Antibiotics* **1975,** 28, 1006-1008
17. Jeffrey, A. M.; Knight, M. ; Evans, W. C. *Biochem. J.* **1972,** 130, 373-381

Chapter 14

Soy Isoflavones in Foods: Database Development

Patricia A. Murphy, Kobita Barua, and Tongtong Song

Department of Food Science and Human Nutrition, Iowa State University, Ames, IA 50011

Although there is considerable interest in plant phytochemicals and their potential health protective effects, there is very limited information on the concentrations of many of these components in foods. Additionally, many clinicians erroneously assume that soy protein is a homogenous material like casein. Recoveries of both internal (4-hydroxybenzyl-2,4,6-trihydroxyphenyl ketone), and external (daidzein, genistein and genistin) standards in 5 different soyfoods, weekly, were evaluated. For precision, soybeans and soymilk were analyzed for within day and for day-to-day precision, bimonthly. Levels of 12 isoflavones (genistein, daidzein and glycitein; glucosides, malonylglucosides, and acetylglucosides) and their forms change during processing. The glucoside forms of the isoflavones are almost 2X the molecular weight of the aglycones. The reported isoflavone levels should be normalized to the aglycone mass 9isoflavanoid equivalent) rather than a simple sum of all isomers. The levels of isoflavones in soybeans, soy-based infant formula, soy flours, isolates, concentrates and TVP from our database will be reported.

Phytochemicals are reported to have a number of health protective effects. However, our knowledge of the concentrations of these components in foods is quite limited. Isoflavones from soybeans, and several other legumes, are classified as phytoestrogens in many literature citations due to their weak estrogenic activity in mammalian systems. A number of epidemiological studies have suggested that consumption of soybeans and soy foods is associated with lowered risks for several cancers (including breast, prostate, and colon)(1), cardiovascular diseases (2,3), and bone health (4). The mechanism(s) for these effects remains to be delineated. Current citations for phytoestrogens in the literature for 1997 exceed 500. Therefore, there is major interest in studying the biological effects of phytoestrogens. In contrast, our knowledge of the levels of these components in foods is limited.

Fortunately, the estrogenic isoflavones are confined to a few plant foods consumed by humans. The major source of phytoestrogens in human diets is isoflavones in soybeans, and soy foods. Estrogenic isoflavones are found in alfalfa and clover seeds, usually consumed as sprouts, chick peas or garbanzo beans, and some pulses. The estrogenic isoflavones found in the other legumes are biochanin A, formononetin, and coumesterol. Our development of a database on isoflavone levels in human foods has been simplified by segregation of isoflavones into a few food plant species.

Soy Foods

Soy ingredients are the major source of isoflavones used in a variety of human clinical, and in animal studies, to determine the mechanism(s) for isoflavones' health protective effects. Unfortunately, some clinicians erroneously believe that all soy protein sources are equal with respect to isoflavone content, and use it as casein is used in nutritional studies. However, isoflavone contents of soy ingredients and foods depend on a number of factors, including the variety of the soybean, crop year, and on the type of processing used to produce the ingredients. We have divided soy foods into three classes: soy ingredients, traditional soy foods, and second generation soy foods(5). Soy ingredients include raw (or unprocessed) soybeans, soy flours (defatted and full-fat), soy concentrates, soy isolates, and texturized soy protein (TVP). Traditional soy foods include soymilks, soy-based infant formulas, tofu, tempeh, natto, miso, and other soy-based foods traditional to Asian cuisine. Second-generation soy foods include items such as soy-based burgers, hot dogs, chicken and bacon analogs, and soy cheeses.

Soy Ingredients Soy ingredients contain the highest concentrations of isoflavones. Defatted soy flours contain the highest concentrations of isoflavones in the ingredient class. The isoflavones tend to associate with the protein fraction in soybeans. There are no detectable isoflavones in soybean oils. Full-fat soy flours have lower isoflavone levels compared to defatted soy flours, due to the dilution of the lipids in full-fat flours. Full-fat flours are a limited section of the commercial marketplace due to their instability due to rapid lipid oxidation. Soy isolates are produced as the alkaline soluble, acid insoluble protein fractions from defatted soy flours to yield approximately 90% protein. The aqueous processing steps result in losses of isoflavones in discarded fractions (6). Soy concentrates are prepared from defatted soy flours by either aqueous or alcoholic extractions. The type of extractant has a profound effect on the isoflavone content, due to the solubility of isoflavones in alcohols. TVP's are prepared by extrusion from defatted soy flours, concentrates or isolates. The remaining isoflavone contents will reflect the concentrations of the starting materials, but the processing history may be unknown by the user of these products.

Database Development and Quality Control

To provide a functional database for clinicians, dieticians, food scientists, and consumers, we attempted to analyze the major soy ingredients in the U.S. food supply for isoflavones. Ingredients selected were representative of: varieties of soybeans grown in Iowa in 1992 and 1995, soy flours, soy isolates, soy concentrates, TVP from the major soy ingredient producers, and soy-based infant formulas.

Analysis of isoflavones was performed by HPLC separation, and quantification by photodiode array detection of the 12 isoflavones found in soy products (Figure 1), and of biochanin A, fomononetin and coumesterol in alfalfa and clover sprouts (Song, *et al*, *Amer. J. Clin. Nutri.*, in press).

Routine quality control is critical to developing a valid database 97). Quality control measures were routinely performed throughout the isoflavone analysis period. These included: analysis of isoflavone standards every day that the samples were run (Figure 2), accuracy estimations by recovery of external and internal standards in 5 soy food matrices monthly (Table I), and precision estimation by evaluation of a coefficient of variation for 4 food matrices for within-day and between-day variance, bimonthly (Table II).

In Figure 2, the importance of this quality control check is evident. Running daily standards each time the samples are analyzed is done to confirm that the HPLC system is operating correctly. The first positive deviation from the mean around the date of 2/19/97 showed that the autosampler was failing. Any samples run on these days were rerun after the instrumentation problem was corrected. The next deviation from the mean, both negative and positive, were on days where the laboratory room heating failed, and then exceeded column heater temperature. Although the latter deviation from our standards mean would be evident to anyone working in the laboratory, the former was not obvious without this internal quality control measure.

The recovery of external standards, genistin, genistein and daidzein, were reasonable (Table I). Recoveries were always higher and more consistent (lower coefficients of variation) for the glucoside, genistin, than the aglycones, genistein and daidzein. These recovery differences were anticipated, since our extraction scheme is optimized to extract the predominant forms of isoflavones in most soy matrices, the glucosides. The glucoside forms, malonylglucoside, acetylglucoside and underivitized glucoside, account for > 95% of the isoflavonoid forms in soy-based foods, except for foods that are highly fermented (such as miso and soy sauce). Recovery of our internal standard, THB, was almost as consistent as genistin. Multiple recovery levels of both internal and external standards were linear over the range of concentrations found in soybeans and soy foods (Song, *et al. Amer. J. Clin. Nurti.*, in press).

Table II presents an example of a bimonthly precision estimate for the isoflavone analysis of one soy matrix stored over the lifetime of the project. No differences were seen in the precision between the two soy matrices, soy flour and freeze-dried soymilk, stored at room temperature and at -29°C, over the lifetime of the project. Additionally, no changes were observed in isoflavone form distribution during storage at room temperature, nor at -29°C (data not shown). Excellent precision was observed for all within-day estimates. Only glycitin deviated outside

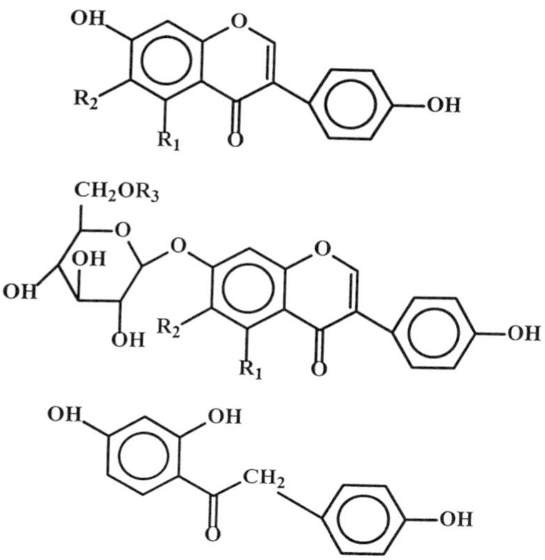

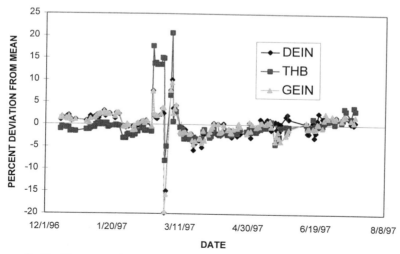

Figure 1. Structures of isoflavones and internal standard, 2,4,4'-trihydroxybenzoin (THB). Daidzein if R_1 & R_2=H; Genistein if R_1=OH & R_2=H; Glycitein if R_1=H & R_2= OCH_3; Glucoside if R_3=H; Acetylglucoside if R_3=C(O)CH$_3$; Malonylglucoside if R_3=C(O)CH$_2$C(O)OH.

Figure 2. Quality control analysis of standards in routine HPLC analysis of soybean isoflavones. DEIN=daidzein; THB=2,4,4'-trihydroxybenzoin; GEIN=genistein.

Table I. Recovery of Genistin, Genistein, Daidzein
And 2,4,4'-Trihydroxybenzioin (THB) in Soy Foods

Food	Recovery				Analyte			
	Daidzein		Genistein		Genistin		THB	
	$X + sd^a$	CV^b	$X + sd$	CV	$X + sd$	CV	$X + sd$	CV
n	25		25		8		21	
TVP	90±9	10	91±6	7	98±6	6	94±5	5
Soymilk	88±8	9	90±11	13	99±7	7	95±8	8
Soybean	81±16	20	95±10	10	99±4	4	98±5	5
Tofu	92±8	8	91±9	10	94±4	4	95±5	5
Tempeh	90±10	11	87±17	19	96±6	6	95±4	4

[a] X = average, sd = standard deviation   [b] CV = coefficient of variation

our 8% CV limits in this example. Since glycitin accounts for <3% of the total isoflavones in this matrix, this deviation was judged to be acceptable.

Data Compilation.   Data compilation was presented two ways. Individual isoflavone contents were measured for all 12 forms found in soy. Since isoflavones are absorbed as the aglycone, the total concentration of isoflavones in food products cannot be expressed as the arithmetic sum of the individual forms. The molecular weight of the glucosides is 1.6 to 1.9 greater than the aglycone. Thus, simple addition of the mass of the isoflavone contents will overestimate the true isoflavone content of the product. Ideally, the molar concentration of isoflavones could be used. However, since many soy products are used by consumers and others not familiar with scientific units, mass concentration units (mg/g and µg/g) are already used on some retail soy products. Therefore, we recommend that total isoflavone contents be adjusted for the molecular weight differences of the glycoside moiety prior to addition.

For example, total genistein in a product should be the sum of genistein + (genistin/1.60) + (malonylgenistin/1.92) + (acetylgenistin/1.76). Total isoflavones can be the sum of the adjusted sums of total genistein + total daidzein + total glycitein. Figure 3 represents the difference in isoflavones reported on an "as is" basis, the sum of mass of each form in the 3 foods, and on a "normalized" basis where the molecular weight differences adjustment has been made to correctly reflect the free isoflavone concentration of the products. It is obvious that failure to normalize the isoflavone data overestimates the true free, or effective isoflavone concentrations of a food source. All subsequent figures in this paper will report only normalized isoflavone totals for daidzein, genistein, and glycitein.

Table II. Within-day and Between-day Precision
Of Isoflavone Analysis[a]

Isoflavone	Mean (μg/g)	Coefficient of Variation (%)	
		Within Day[b]	Between Day[c]
Daidzin	188±6	3.3	3.1
Glycitin	52±5	3.0	9.5
Genistin	223±10	3.1	4.4
Malonyldaidzin	1,430±29	2.6	2.0
Malonylglycitin	131±3	5.6	2.5
Malonylgenistin	1,379±26	2.1	1.9
Acetylgenistin[d]	28±1	1.4	3.5
Daidzein	21±1	1.8	5.0
Genistein	19±1	1.9	4.8
Total Daidzein	860±17	2.6	1.9
Total Glycitein	103±4	4.7	4.1
Total Genistein	893±18	2.1	2.1

[a] Soybean flour stored at -29°C over 24 months. [b] n=4. [c] n=5. [d] acetyldaidzin, acetylglycitin, and glycitein not detected.

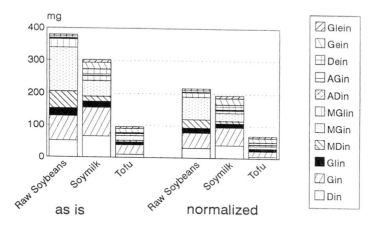

Figure 3. Isoflavones in soybeans, soymilk and tofu expressed on an "as is" basis and on a "normalized" basis. Din=daidzin; Gin=genistin; Glin=glycitin; MDin=malonyldaidzin; MGin=malonylgenistin; MGlin=malonylglycitin; ADin=acetyldaidzin; AGin=acetylgenistin; AGlin=acetylglycitin; Dein=daidzein; Gein=genistein; Glein=glycitein.

Isoflavones in Soy Ingredients

The results of our database development for soy ingredients and soy-based infant formulas are represented in Figures 4-9. In Figure 4, the yearly variability in total isoflavone levels in soybeans, which are destined for ingredient manufacture as well as soybean oil production, is evident. In the 1992 soybeans, the range of total isoflavones was from 800 to 1500 μg/g. In the 1995 soybeans, the range was from 900 to 2050 μg/g. In the 1996 soybean, the range was from 900 to 3250 μg/g. This variability represents both season differences in isoflavone expression and genetics of the soybean variety. Farmers plant different varieties of soybeans each year as seed companies make agronomic improvements and sell different varieties. We have observed up to 5 fold differences in isoflavone concentrations in single varieties grown in different locations or crop years (8). It is our opinion that environment affects on isoflavone concentrations are much greater than genetics. We have well controlled agronomic studies that support this opinion (Fehr, Hoeck, Murphy, unpublished).

Figure 5 presents the levels of isoflavones found in commercial soy flours. Soy flours may be prepared from soybeans or hulled soybeans. The hull contributes a very small fraction to isoflavone totals (6). Commercial soy flours may be full fat, but, due to stability issues, most commercial flours will be defatted. The range in isoflavones observed in these commercial flours, mimics the variation seen in intact soybeans.

Commercial soy concentrates can be divided into two types ranging in isoflavone concentrations from 50 to 1000 μg/g (Figure 6). Prior to 1994, almost all commercial soy concentrates could be represented by samples A-f, the very low isoflavone containing concentrates. These low values reflect the popular technology for preparing concentrates, by ethanol washing to remove oligosaccharide and undesirable flavors, and to retain the functional properties desirable in a concentrate. Since 1994, soy ingredient manufacturers have recognized the marketing advantage to having some soy concentrates with isoflavone levels closer to that of intact soybeans or soy flours. This has resulted in concentrates available with isoflavone levels similar to flours. Both types of products are currently available in the marketplace. the ways to determine what type of concentrate one has, are to analyze for isoflavones, or require the manufacturer to document isoflavone levels in the product. As a general rule, alcohol extracted concentrates can be expected to have low isoflavone levels, unless the manufacturer has added the isoflavone fraction back to the product.

Soy isolates are generally the most expensive soy ingredient and contain isoflavones at one third to one half found in intact soybeans ranging from 470 to 1050 μg/g (Figure 7). The aqueous alkali and acid steps used to produce isolate results in losses of isoflavones to discarded fractions in product production (6). Some manufacturers are recovering this now valuable co-product in their isolate manufacturing operations.

Four commercial TVP isoflavone concentrations are presented in Figure 8. TVPs' isoflavone levels seem to fit a bimodal population, as concentrations do. This reflects the variety of soy ingredients that can be used to produce TVPs. TVPs can

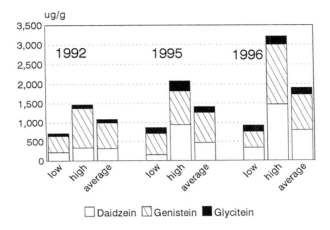

Figure 4. Isoflavones in soybeans harvested in 1992, 1995 and 1996. Aglycone total concentrations or "normalized" concentrations of daidzein, genistein and glycitein are shown.

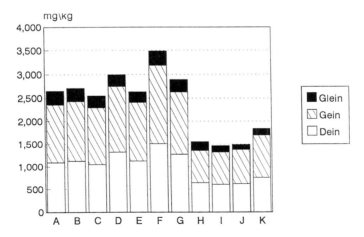

Figure 5. Isoflavones in commercial soy flours. Aglycone total concentrations or "normalized" concentrations of daidzein, genistein and glycitein are shown.

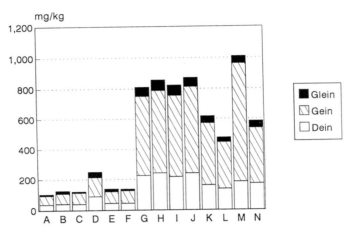

Figure 6. Isoflavones in soy concentrates. Aglycone total concentrations or "normalized" concentrations of daidzein, genistein and glycitein are shown.

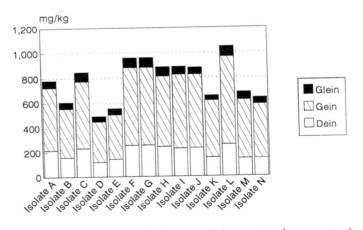

Figure 7. Isoflavones in soy isolates. Aglycone total concentrations or "normalized" concentrations of daidzein, genistein and glycitein are shown.

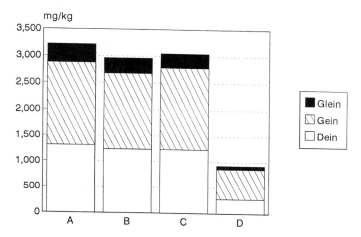

Figure 8. Isoflavones in texturized vegetable (soy) protein. Aglycone total concentrations or "normalized" concentrations of daidzein, genistein and glycitein are shown.

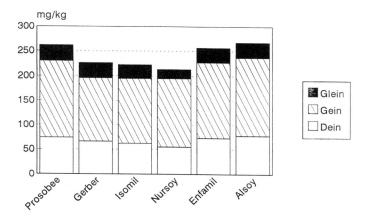

Figure 9. Isoflavones in soy-based infant formulas. Aglycone total concentrations or "normalized" concentrations of daidzein, genistein and glycitein are shown.

be made from soy flours, soy concentrates, and soy isolates (9). TVPs may be produced by extrusion of semi-moist soy ingredients, spinning of soy fibers from very concentrated soy protein solutions, or other technologies to produce solid materials resembling muscle proteins. The end user may have no knowledge of the initial soy ingredient used to produce a TVP. Either an isoflavone measurement must be made, or the TVP supplier must provide this information. The isoflavones in the TVP presented in Figure 8 appear to be made from flours for the high isoflavone containing products, and from isolate, for the product containing 950 μg/g.

Figure 9 presents the isoflavone levels found in commercial soy-based infant formula available in the United States. These concentrations are on a dry formula weight basis. Reconstituted formula would contain 25-30 μg/ml if prepared according to the manufacturers directions. Soy-based infant formulas are produced using soy isolate as the soy source. Although there were some statistically significant differences in individual isoflavone amounts in these commercial formulas, they are surprisingly similar in comparison, with the difference in isoflavone concentrations observed in soybeans and other soy products. When isoflavone levels in these formulas are adjusted for the percent soy isolate reported on the label on the product formulation, the differences between products is even smaller.

In summary, isoflavone database development requires attention to analytical quality control to insure the goodness of the data destined for the database. A variety of quality control measures should be employed, including precision and accuracy measures of the matrix effects, and quality control of the instrumentation. Data must be expressed correctly on a mass basis, taking into account the differences in the molecular weight of the isoflavanoid forms. Soy ingredients vary according to the variety, level of processing, and type of processing. Ranges of isoflavone concentrations are known, but exact levels for a particular product should be measured by the investigator, or supplier.

Acknowledgements. This work was supported in part by the U.S. Army Medical Branch and Material Command under DAMD17-MM 4529EVM, and by the Iowa Agricultural and Home Economics Experiments Stations, and published as Journal No. J-17657, project 3353.

Literature Cited

1. Messina, M.; Barnes, S. *J. Natl. Cancer Inst.* **1991**, 83, 541-546.
2. Anderson, J.W.; Johnstone, B.M.; Cook-Newell, M.E. *New Engl. J. Med.* **1995**, 333, 276-283.
3. Anthony, M.S.; Clarkson, T.B.; Hughes, C.L.; Morgan, T.M.; Burke, G.L. *J. Nutri.* **1996**, 126, 43-50.
4. Burham, H.A.; Alekel, L.; Hollis, B.W.; Amin, D.; Stacewicz-Sapuntzakis, M.; Guo, P.; Kukreja, S.C. *J. Nutri.* **1996**, 126, 161-167.
5. Wang, H.J.; Murphy, P.A. *J. Agric. Food Chem.* **1994a**, 42,1666-1673.
6. Wang, H.J.; Murphy, P.A. *J. Agric. Food Chem.* **1996**, 44, 2377-2383.

7. Mangels, A.R.: Holden, J.M.; Beecher, G.R. et al. *J. Amer. Diet. Assoc.* **1993**, 93, 284-296.
8. Wang, H.J.; Murphy, P.A. *J. Agric. Food Chem.* **1994b**, 42, 1674-1677.
9. Snyder, H.E.; Kwon, T.W. In *Soybean Utilization*, Von Nostran Reinhold Co., New York, NY, 1987, pp. 135-142.

Chapter 15

Human Bioavailability of Soy Bean Isoflavones: Influences of Diet, Dose, Time, and Gut Microflora

S. Hendrich, G.-J. Wang, X. Xu, B. -Y. Tew, H.-J. Wang, and P. A. Murphy

Department of Food Science and Human Nutrition, Iowa State University, Ames, IA 50011

To characterize the bioavailability of the anticarcinogenic soy isoflavones, men and women were fed soy foods containing 0-2.7 mg isoflavones/kg. Effects of background diet, soy food type, and dose were examined in studies using randomized cross-over designs. Plasma, urine and fecal isoflavones were measured by reverse-phase HPLC. After a single dose of soy, plasma daidzein and genistein were similar. Urinary recoveries of daidzein over 24 h from various soy foods fed within varying background diets were 25-50%, and of genistein, 10-20% of the dose given. Daidzein recovery was significantly greater than that of genistein. Isoflavone recovery did not depend upon soy food or background diet, and isoflavone excretion increased linearly with dose. Human gut microfloral breakdown of isoflavones was inversely related to plasma and urinary isoflavone content, with three distinct phenotypes (high, moderate, and low fecal isoflavone excretor) evident and likely to significantly influence isoflavone bioavailability.

Isoflavones are proposed anticarcinogens, found in soy beans in amounts of 1-3 mg/g (1) and in soy foods in amounts of 0.025--3 mg/g (2). Only trace amounts of isoflavones are found in other legumes, making soy a unique source of these potentially beneficial food components. Isoflavones are antioxidants in vitro and in vivo (3, 4). Their weak estrogenicity may be related to both toxic and potentially beneficial effects (5). Isoflavones are tyrosine kinase and DNA topoisomerase II inhibitors (6, 7). They also inhibit angiogenesis in an in vitro model (8). Any or all of these functions might account for the anticarcinogenic effects of isoflavones. Isoflavone extracts made from toasted, defatted soy flakes in amounts of 1 mmol total isoflavone per kg of diet are anticarcinogens at the stage of promotion in rat hepatocarcinogenesis models (9, 10). The anticarcinogenic ability of purified isoflavones remains to be proven, as well as their mechanisms of action in vivo.

The ability of isoflavones to exert such anticarcinogenic effects depends upon their bioavailability, defined in a nutritional sense as the relative ability of these compounds to reach their sites of action in biologically active forms. Bioavailability of dietary compounds depends upon their structure, dose, and characteristics of their food matrices. In many cases, interactions of the compounds with gut microflora may also be significant in determining the compounds' biological effects.

Influence of Soy Food Type and Isoflavone Structure on Isoflavone Bioavailability

Soy isoflavone structure determines some aspects of the bioavailability of these compounds (Figure 1). Isoflavones in most soy foods are more than 90% in glycosidic form, which seemingly must be cleaved by gut microfloral β-glycosidases before the isoflavone aglycones can be absorbed, because those glycosides are probably poor substrates for mammalian β-glycosidases (11). Isoflavone aglycones, which are present in fermented soy foods in relatively great quantities (2), may be somewhat more rapidly absorbed than isoflavone glycosides (12). Plasma and urinary isoflavones were similar in rats fed genistein or an equimolar amount of genistein glycoside from soy extract at 10-50 h after a single dose in rats (12), suggesting good efficiency of glycoside cleavage. Thus, whether dietary isoflavones are present as glycones or aglycones may not be important to their availability. The presence of malonyl or acetyl side chains on the glycosides may also be unimportant with respect to isoflavone absorption from soy foods, because when women were fed the same single dose of isoflavones from tofu (relatively high in glycones and acetyl glycones) v. texturized vegetable protein (TVP, relatively high malonyl glycone content), urinary and plasma isoflavones did not differ with soy food (13). Urinary excretion of genistein was greater after tofu feeding than after TVP feeding only because the total genistein content of the tofu was greater than that of TVP.

The phenolic structure of isoflavones seems to be a major determinant of their availability because such hydroxyls are readily metabolized to glucuronide conjugates, which are excreted rapidly in bile and urine (14). Biliary excretion is likely the main limiting factor with respect to the dose percentage of isoflavones that is systemically available. The generally low recovery of isoflavones from feces, and recovery of a total of only 10-50% of ingested isoflavones in urine (12, 13, 15-17) suggests that biliary excretion of isoflavones is followed by their breakdown, catalyzed by gut microorganisms (17).

Influence of Dose and Time Course on Isoflavone Bioavailability

A series of human feeding studies has been performed to assess the influence of isoflavone dose on isoflavone excretion and plasma concentrations. In the first study, 8 young women were fed a single dose at breakfast of 8.0 μmol total

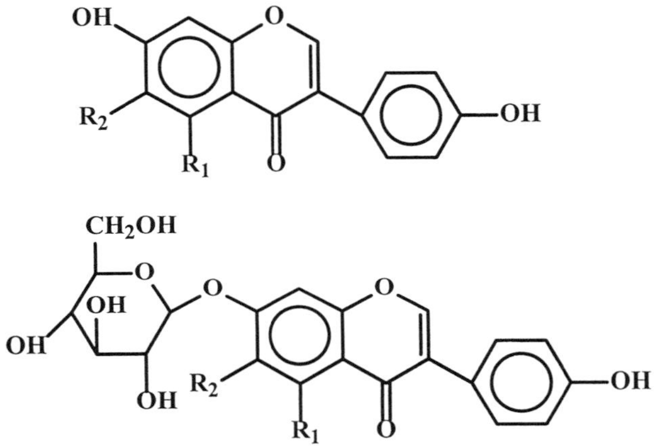

Figure 1. Structures of soy bean isoflavones. For daidzein, $R_1=R_2=$ H, for genistein, R_1 = OH and for glycetein, R_1=H and R_2=OCH$_3$. The respective glucosidic forms are daidzin, genistin and glycetin. The three isoflavones may also occur in foods as malonated and acetylated glucosides.

isoflavones from soy milk/kg body weight within a background diet consisting of 3 meals of chocolate-flavored Carnation Instant Breakfast in skim milk. The peak excretion of isoflavones occurred between 5-11 h after dosing, with excretion falling to nearly nil at 24 h after dosing. Urinary excretion of daidzein was more than twofold greater than that of genistein, although the dose of daidzein was only 25% greater than was the ingested dose of genistein. Plasma daidzein averaged 6.84 μmol/L and genistein 3.3 μmol/L at 6.5 h after soy milk dose (18). A second study feeding single doses of soy milk isoflavones of 2.0, 4.0 and 8.0 μmol/kg to 12 women (age 24 $\pm$ 6 y (mean $\pm$ standard deviation) in a randomized cross-over design showed similar results of a statistically significant twofold difference in relative urinary recovery of daidzein v. genistein at all doses (16). Dose was linearly related to plasma isoflavone level at 6.5 h after dosing (e.g., 0.7, 1.1 and 2.1 μmol genistein/L corresponding to doses of 2.0, 4.0 and 8.0 μmol/kg). Dose was similarly related to urinary isoflavone levels (24, 58 and 79 μmol total isoflavone excreted at 2.0, 4.0 and 8.0 μmol/kg doses, respectively).

A subsequent randomized cross-over design study examined 7 women fed three isoflavone-containing meals per day at 5 h intervals for a total daily dose of 10, 20 and 30 μmol isoflavones/kg. Peak urinary isoflavone excretion was prolonged, with nearly the same total amounts excreted between 0-12 and 12-24 h after the first isoflavone dose, compared with significantly decreased urinary isoflavone levels at 12-24 h v. 0-12 h after dose in the single dose study (16). But total urinary isoflavone excretion was still linearly related to isoflavone dose (88, 159 and 250 μmol excreted, respectively, at doses of 10, 20 and 30 μmol/kg) (17). For 5 of 7 subjects, urinary recovery of daidzein was significantly greater than that of genistein. But two subjects, who excreted significantly greater amounts of total isoflavones than did the other 5 subjects, showed about 34% recovery of both ingested genistein and daidzein in urine. The two- to threefold greater urinary isoflavone recovery in these two subjects was accompanied by tenfold greater recovery of isoflavones in their feces compared with other subjects.

To characterize longer term dose effects on bioavailability of isoflavones, eight women and eight men, ages 20-28, were fed 0, 3.0 or 6.0 μmol isoflavones/kg from soymilk in breakfast for seven days in a randomized cross-over design. Each subject consumed an ad libitum diet with the soymilk with a one-week washout period between soymilk doses. Twenty-four h urine collections were conducted on the last two days of each feeding period and isoflavones measured by reverse-phase HPLC as described (16). In all subjects, urinary excretion of isoflavones was nearly the same on days 6 and 7. Urinary daidzein was twofold greater than genistein and tenfold greater than equol on both feeding days. Total 24 h urinary isoflavone excretion was 48 and 96 μmol at 3.0 and 6.0 μmol/kg doses, respectively in women (25% of daily dose). Men excreted an average of 80 and 140 μmol total isoflavones over 24 h at doses of 3.0 and 6.0 μmol/kg, respectively (30-34% of daily dose). These data suggest that men have greater bioavailability of isoflavones than do women. Excretion of daidzein, genistein and equol seemed to reach a steady state by six days of feeding in these subjects, with a relatively linear dose/response.

In general, human bioavailability of isoflavones is linearly related to their dose within a broad dose range (2-30 μmol/kg body weight). Daidzein is more bioavailable than is genistein, when total urinary excretion is used as an indicator. Isoflavones from a single dose are usually cleared from the body within 24 h. More frequent doses prolong clearance time. Moderate gender differences in isoflavone bioavailability seem to occur, at least in young adults.

Isoflavone Metabolism: Influence of Gut Microflora

The consistently low recovery of isoflavones from feces in human bioavailability studies (13, 15-17) coupled with urinary recovery of only 10-50% of ingested isoflavones indicates either significant disappearance or significant storage of these compounds. Storage of isoflavones within the the human body seems highly unlikely. At present, there exists no evidence for such a phenomenon. Furthermore, the chemical nature of these compounds compels their rapid biliary and urinary excretion after glucuronidation. The fate of isoflavones after biliary excretion has not been well-characterized chemically, but flavonoids and isoflavonoids may be degraded by strains of *Clostridium*, *Butyrivibrio* and *Eubacterium* found in the guts of rats, cattle and humans (19-21). The half-lifes of disappearance of isoflavones in an *in vitro* anaerobic incubation of a suspension of human feces in brain/heart infusion (BHI) media were 7.5 and 3.3 h for daidzein and genistein, respectively (17). Genistein may be more susceptible to C-ring cleavage by virtue of its 5-OH group, a moiety absent from daidzein (22).

Some individuals, who have much greater fecal isoflavone excretion than most subjects, experience more prolonged and greater isoflavone bioavailability, with similar bioavailability of daidzein and genistein, in contrast with the majority of subjects (17). This suggests that the presence or absence of certain gut microfloral species may be an important determinant of isoflavone bioavailability.

Gut microfloral metabolism of isoflavones in vitro has been studied in 15-20 subjects over a ten-month period. Adults ages 21-45, who had not taken antibiotics for 3 months before or during the study period, deposited fecal samples, which were immediately homogenized as tenfold (weight/volume) dilutions in BHI media. After addition of 590 μmol/L each of daidzein and genistein (23), the samples were incubated anaerobically for 48 h. The BHI medium contained cysteine hydrochloride (0.5 mg/L) as a reducing agent, and resazurine (1.0 mg/L) as an oxygen indicator. Media samples were analyzed for isoflavone content at 0, 6, 12, 24 and 48 h by HPLC (16). The rate constant, k, of degradation of daidzein, calculated as the negative slope of the regression line plotted for isoflavone content of each sample over time, was significantly different at day 0 of the study among subjects grouped according to three fecal excretor phenotypes: Low; k = 0.012 (n = 5), Moderate; k = 0.055 (n = 10) and High; k = 0.299 (n = 5). At day 300, Low (n = 5) and Moderate (n = 4) phenotypes had similar daidzein degradation rate constants of 0.053 and 0.073, respectively, and differed significantly from the daidzein degradation rate constant of the High excretor phenotype (n = 5), k = 0.326. At day 0, genistein degradation rate constants were 0.023, 0.163 and 0.299 for Low, Moderate and High excretor phenotypes, respectively, each group significantly different from the other groups (p < 0.05). At day 300, genistein degradation rate constants still differed significantly among the three phenotypes, k = 0.049 (Low), 0.233

(Moderate) and 0.400 (High). These data suggest relatively stable human gut microfloral differences in ability to degrade isoflavones. The microorganisms responsible for these differences remain to be determined. It is likely that such differences would influence the biological effects of isoflavones, given our previous observations (17). Further characterization of such differences and development of screening methods may be important for validation of isoflavone intakes in human clinical trials of the health effects of soy.

Conclusions

Isoflavone bioavailability depends upon the phenolic nature of these compounds, which determines their major metabolic products. Gut microflora also exert profound effects on isoflavone availability, with microfloral breakdown of these compounds limiting their reabsorption and systemic availability. The presence of varying glycosidic and aglycone forms of isoflavones in soy foods probably has little impact on the bioavailability of these compounds, which is linearly related to their dose over a broad dose range.

Acknowledgments

This work was supported by National Institutes of Health grant CA-56308-02 and the Center for Designing Foods to Improve Nutrition, Iowa State University.

Literature Cited

1. Wang, H.-J.; Murphy, P.A. *J. Agric. Food Chem.* **1994,** *42,* 1674-1681.
2. Wang, H.-J.; Murphy, P.A. *J. Agric. Food Chem.* **1994,** *42,* 1666-1673.
3. Naim, M.; Gestetner, B.; Bondi, A.; Birk, Y. *J. Agric. Food Chem.* **1976,** *24,* 1174-1177.
4. Wei, H.; Wei, L.; Frenkel, K.; Bowen, R.; Barnes, S. *Nutr. Cancer* **1993,** *20,* 1-12.
5. Farmakalidis, E.; Hathcock, J. N.; Murphy, P.A. *Food Chem. Toxicol.* **1985,** 23, 841-745.
6. Akiyama, T.; Ishida, J.; Nakagawa, S.; Ogawara, H.; Watanabe, S.; Itoh, N.; Shibuya, M.; Fukami, Y. *J. Biol. Chem.* **1987,** *262,* 5592-5595.
7. Okura, A.; Arakawa, H.; Oka, H.; Yoshinari, T.; Monden, Y. *Biochem. Biophys. Res. Comm.* **1988,** *157,* 183-189.
8. Fotsis, T.; Pepper, M.; Adlercreutz, H.; Flerischmann, G.; Hase, T.; Montesano, R.; Schweigerer, L. *Proc. Natl. Acad. Sci. U.S.A.* **1993,** *90,* 1690-1694.
9. Lee, K.-W.; Wang, H.-J.; Murphy, P.A.; Hendrich, S. *Nutr. Cancer* **1995,** *24,* 267-278.
10. Hendrich, S.; Lu, Z.; Wang, H.-J.; Hopmans, E. C.; Murphy, P. A. *Amer. J. Clin. Nutr. (suppl.), in press.*
11. Brown, J. P. In *Role of the Gut Flora in Toxicity and Cancer,* Rowland, I.R., Ed., Academic Press: San Diego, CA, 1988; pp 109-144.
12. King, R. A.; Broadbent, J. L.; Head, R.. J. *J. Nutr.* **1996,** *126,* 176-182.

13. Tew, B.-Y.; Xu, X.; Wang, H.-J.; Murphy, P.A.; Hendrich, S. *J. Nutr.* **1996,** *126,* 871-877.
14. Sfakianos J.; Coward, L.; Kirk, M.; Barnes, S. *J. Nutr.* **1997,** *127,* 1260-1266.
15. Hutchins, A. M.; Slavin, J. L., Lampe, J. W. *J. Am. Diet. Assoc.* **1995,** *95,* 545-551.
16. Xu, X.; Wang, H.-J.; Murphy, P. A.; Cook, L., Hendrich, S. *J. Nutr.* **1994,** *124,* 825-832.
17. Xu, X.; Harris, K. H.; Wang, H.-J.; Murphy, P. A.; Hendrich, S. *J. Nutr.* **1995,** *125,* 2307-2315.
18. Xu, X.; Wang, H.-J.; Murphy, P. A.; Cook, L. R.; Hendrich, S. In *Natural Protectants Against Natural Toxicants,* Bidlack, W. R.; Omaye, S. T., Eds., Technomic Publishing Co., Inc., Lancaster, PA, 1995, pp. 51-58.
19. Cheng, K.-J.; Jones, G. A.; Simpson, F. J.; Bryant, M. P. *Can. J. Microbiol.* **1969,** *15,* 1365-1371.
20. Krishnamurty, H. G.; Cheng, K.-J.; Jones, G. A.; Simpson, F. J.; Aatkin, J. E. *Can. J. Microbiol.* **1970,** *16,* 759-767.
21. Winter, J.; Popoff, M. R.; Grimont, P.; Bokkenheuser, V. D. *Intl. J. System. Bacteriol.* **1991,** *41,* 355-357.
22. Griffiths, L. A.; Smith, G. E. *Biochem. J.* **1972,** *128,* 901-911.
23. Song, T.; Murphy, P. A. *Am. J. Clin. Nutr. (suppl.)*, *in press.*

Chapter 16

The Color Stability and Antioxidative Activity of an Anthocyanin and γ-Cyclodextrin Complex

Hirotoshi Tamura, Mitsuyoshi Takada, Atsushi Yamagami, and Kohei Shiomi

Department of Bioresource Science, Kagawa University, Miki-cho, Kita-gun, Kagawa 761-07, Japan

Since the malvidin type of anthocyanins was found to be the strongest antioxidant among 6 representative types of anthocyanins, which activity was evaluated by TBA test, the utilization of the malvidin type of anthocyanins as food additives was investigated with great interest. However, anthocyanins are apt to change their chemical structure to forms which are irreversibly decomposed to lower molecules in a weakly acidic aqueous solution. For example, the color of anthocyanins possessing the anhydrobase and/or flavylium forms, will rapidly fade away due to the formation of a colorless pseudobase.

The effect of α, β, and γ-cyclodextrins on the color stability of anthocyanins was investigated with respect to the formation of inclusion compounds. The color stability of malvidin 3-(6- O-p-coumaroylglucoside) was greatly enhanced by the formation of an inclusion molecule with γ-cyclodextrin. Furthermore, antioxidative activity of the anthocyanins was retained after formation of the host-guest complex with γ-cyclodextrin at pH 7.4.

Low-density lipoprotein (LDL), which is the major carrier of plasma cholesteryl ester, has been implicated as one of the agents for coronary heart disease (CHD). In many countries, high intake of saturated fats has been strongly correlated with high mortality from CHD. French people, in general, have shown a relatively low frequency of CHD, whereas groups of people in other European countries have shown a higher frequency (1,2). This population difference in the occurrence of CHD has been termed the French Paradox. Further, oxidative modification of the polyunsaturated lipid components of LDL, by active oxygen species, may also have a role in causing atherosclerosis (3). Recently, the French Paradox has been attributed, in part, to a generally higher level of red wine consumption by the French. The wine alcohol itself may be responsible for preventing CHD (4,5). Additionally, non-alcoholic components also appear to have potent and beneficial antioxidant properties with regards to the oxidation of human LDL (2). Specifically, the polyphenols in red wine, such as tannins, catechins and anthocyanins, have received attention as the chemicals potentially responsible for protecting LDL from oxidative modification. The proanthocyanidin type of tannins, catechins, and anthocyanins are the main polyphenols in grape extracts (6-8). Anthocyanins, which are representative flavonoids, and widely found in plants, readily change their chemical form in aqueous solution. Previous researchers have not evaluated the various physiological functions of the individual anthocyanin forms [i.e., flavylium, anhydrobase or pseudobase forms]. However, various physiological functions of anthocyanin equilibrium mixtures (9,10) have been reported.

In general, it is believed that functional flavonoids, including anthocyanins, show physiological activities as the aglycon form in blood vessels. Duke has reported (9) and summarized the biological activity of some anthocyanidins. For instance, pelargonidin has been found to act both as an anti-viral and an anticarcinogenic agent. Moreover, delphinidin was also found to have anticarcinogenic and antioxidative properties. However, analysis of standard solutions of Vaccinium myrtillus anthocyanins (VMA) and rat plasma, after oral administration (400mg/kg), demonstrated that the absorption of anthocyanins occurred in the rat plasma (11). About 5% of anthocyanidin glucosides, not anthocyanidin, were absorbed into the blood. This was the first report that anthocyanin glucosides themselves were observed in blood, demonstrating the important role of anthocyanins in these physiological functions (12). Therefore, the utilization of anthocyanins as food additives has been accorded great interest. However, one of the drawbacks has been that anthocyanins are prone to changes in their chemical structure in weakly acidic aqueous solution. Consequently, coloring species of anthocyanins, which include the anhydrobase and flavylium forms, in a weakly acidic solution, will quickly fade into a colorless pseudobase form, an undesirable effect. Some stabilization mechanisms of food colorants must therefore be considered.

Cyclodextrin can bind to various kinds of low molecule compounds into its cave structure, thereby stabilizing the low molecule organic compounds by the inclusion effect. This includes hydrophobic interaction and steric protection from hydration. Some examples of anthocyanin-cyclodextrin complexes have previously been reported (13). It was observed that adding β-cyclodextrin to solutions of callistephin and chrysanthemin at pH 2.0, and 26°C, resulted in a strong decrease in the flavylium chromophore absorption band. In contrast, malvin, and α- or β-cyclodextrin complexes did not appear to affect the visible absorption spectrum of the flavylium chromphore. This suggests that inclusion of malvin into the cyclodextrin cavity did not occur (14-16). In addition, no evidence for color stabilization, or an effect of the anthocyanin-cyclodextrin complex on antioxidative activity was found. In our objectives, inclusion of colored forms and colorless forms of the anthocyanin into a cyclodextrin structure was expected to help in maintaining the food factor functionality of the anythocyanin. The inclusion effect of cyclodextrin on various kinds of malvidin type of anthocyanins was studied.

Isolation of anthocyanins in grape skins

HPLC analysis. HPLC analysis of Muscat Bailey A pericarps was carried out according to the method described by Goto and co-workers (17) as follows: Column, C18, 5mm Develosil (250 X 4.6 mm i.d.); temperature was controlled at 40°C; detector 280 nm and 520 nm; mobile phase, A: AcOH : CH3CN : H2O : H3PO 4 = 8 :10 : 80.5 : 1.5%; mobile phase, B: AcOH : CH3CN : H2O : H3PO4 = 20 : 25 : 53.5 : 1.5% (v/v) . The flow rate of the eluent was 1 mL/min. The analysis was accomplished by a linear gradient elution from solvent A to solvent B in 30 min. Isolation of anthocyanin pigments from 1.4 kg of Muscat Bailey A pericarps using 1.8 L of 0.5% TFA-50%

MeOH aqueous solution was carried out according to the method described by Tamura et al. (18) . Malvidin 3-glucoside, malvidin 3,5-diglucoside, malvidin 3-O-(6-O-p-coumaroylglucosido)- 5-glucoside, and malvidin 3-O-(6-O-p-coumaroylglucoside) were purified with an ODS column. The purity of each anthocyanin was checked by analytical HPLC at 290 nm and 520 nm. The four pigments with purity greater than 95% were used to assay antioxidative activity and color stability, in a cyclodextrin aqueous solution.

Oxidation system of linoleic acid by ferrous sulfate. The oxidation of linoleic acid was conducted using a modification of a method reported previously (19, 20). Linoleic acid (17.8 mmol) was diluted with 4.9 mL of a trizma-buffer solution (0.25 mM, pH 7.4) containing 0.2% SDS and 0.75 mM potassium chloride. Lipid peroxidation was initiated by adding 1 mmol ferrous sulfate. The antioxidant solution (0.1mL) was added into the buffer solution when necessary. The total volume of the reacting lipid peroxidation solution was then adjusted to 5 mL. Incubation was continued for 16 h at 37°C in the dark. Adding 9.1 mmol BHT alcoholic solution to the tube stopped the reaction.

Antioxidative activity assay (21,22). The antioxidant (5 μmol) was dissolved in 0.5% trifluoroacetic acid-ethanol or 0.5% trifluoroacetic acid-water, then the solution was made up to 5 ml with the same solvent. The diluted antioxidant solution (0.1 mL) was mixed with the lipid peroixidation solution (4.9 mL). The antioxidative activity was quantitatively determined using the thiobarbituric acid (TBA) method (23). The amount of TBA reactive substances was estimated by malondialdehyde formed from 1,1,3,3-tetraethoxypropane.

Color stability of anthocyanins. The color stability of malvidin 3,5-diglucoside, malvidin 3-glucoside and their p-coumaroyl ester was measured at pH 3, 5 and 7.0, and 30°C for 24 h. The decrease in visible maximal absorption (540 nm for pH 3.0, 550nm for pH 5.0, 580nm for pH 7.0) of the anthocyanins was recorded against time course. UV-VIS spectra were measured with a JASCO V-520SR spectrophotometer. The pH solutions mentioned above were prepared by mixing 0.1M of a citric acid aqueous solution with 0.2 M of a Na_2HPO_4 aqueous solutions (McIlvain buffer solution).

Antioxidative activity of anthocyanins

Antioxidative activity of the four pigments extracted from Muscat Bailey A grapes was previously evaluated on the basis of the amount of malondialdehyde formed by the autoxidation of linoleic acid in Trizma buffer (pH 7.4) (22). p-Coumaric acid did not affect the antioxidative activity. Malvidin 3-glucoside acted as one of the polyphenol antioxidants. However, the mixture of malvidin 3-glucoside and p-coumaric acid did not show cooperative antioxidative activity. Malvidin 3-glucoside acylated with p-coumaric acid, in contrast, showed the strongest antioxidative activity. p-coumaric acid thus appears to act as an intramolecular synergist. Moreover, the relative antioxidative activity of anthocyanidin among flavonoids was of great interest, and subsequently compared (Fig. 1 and 2).

Each flavonoid collected commonly had ortho diols on the B-ring. Thus, cyanidin was selected as the representative anthocyanidin, and compared with the flavonoids. The antioxidative activity of cyanidin was found to be the same as that of catechin and quercetin. Luteolin, a flavone, was the strongest antioxidant. The antioxidative activity of six representative anthocyanidin 3,5-diglucosides, and their p-

Control
BHT
α - Tocopherol
Luteolin
Eriodictyol
Taxifolin
(+)-Catechin
Cyanidin
Quercetin
antioxidant(10μM)
0
5
10
15
20
Amount of MDA (nmol/mg linoleic acid)

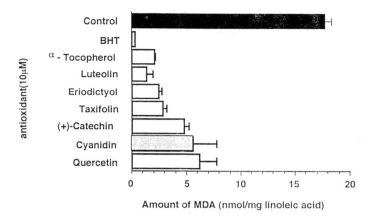

Figure 1. Comparison of antioxidative activity between cyanidin and the other flavonoids in the linoleic acid system at pH 7.4. Lipid peroxidation was induced by 400 μM $FeCl_2$. Reported values are the mean ± standard deviation (n=4).

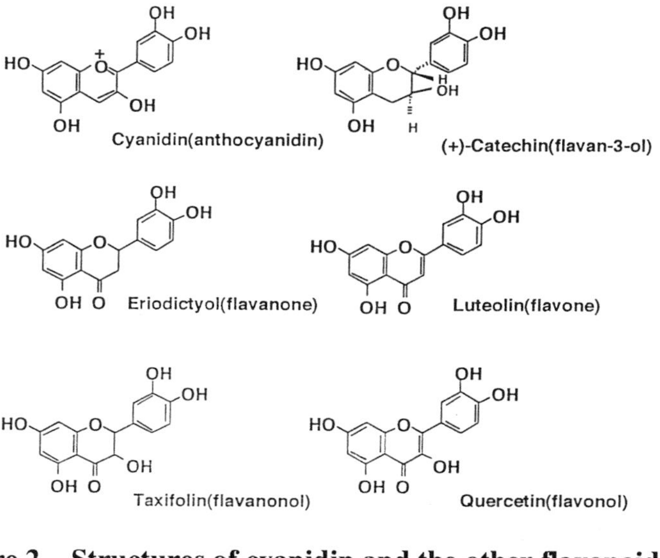

Cyanidin(anthocyanidin)
(+)-Catechin(flavan-3-ol)
Eriodictyol(flavanone)
Luteolin(flavone)
Taxifolin(flavanonol)
Quercetin(flavonol)

Figure 2. Structures of cyanidin and the other flavonoids having orth diol on the B ring.

coumaroyl esters, was also compared to one another (Fig. 3). The malvidin type of anthocyanin was found to be a powerful antioxidant. Antioxidative activity of α-Tocopherol and related phenolic antioxidants has been previously reported in a self-initiated autoxidation system (24) by Burton and Ingold, whereby α-Tocopherol was the strongest antioxidant in the related phenolic compounds. The stability of free radicals formed by the autoxidation was attained by reaction inhibition, due to bulk methyl groups close to the phenoxy radical. Two methoxy groups on the B ring of the malvidin chromophore might participate in the stability of free radicals for the same reason. In addition, all of the acylated anthocyanins also exhibited greater antioxidative properties. In each case, the p-coumaroyl moiety appeared to greatly contribute to antioxidative activities as a synergist. Among the anthocyanins, antioxidative activity of the cyanidin type of anthocyanins was less than that of the malvidin type of anthocyanin. So, the antioxidative potency in wine grapes might be, in large part, due to the main four malvidin series of anthocyanins contained in Muscat Bailey A, Cabernet Sauvignon, and some other grape varieties, more so than the catechins.

Stability of anthocyanins in buffer solutions with cyclodextrin

As the antioxidative activity of the malvidin type of anthocyanins showed the highest values amongst the six representative anthocyanins having different functional groups on the B-ring, color stability and antioxidative activity of the malvidin series of anthocyanins, in buffer solutions with cyclodextrins, were also studied. Cyclodextrin can include various chemicals in the nonpolar cave of the host molecules to form a stable host-guest complex. Thus, α-cyclodextrin (cavity size, 6 Å) can include one molecule of benzene, and β-cyclodextrin (8 Å) can include one molecule of naphthalene as shown in Fig. 4. The larger γ-cyclodextrin (10 Å) can include two molecules of naphthalene (25). On the other hand, the molecular size of anthocyanidins in nature is approximately 6.6 X 12 Å (Fig. 5). Thus, anthocyanin molecules, such as nathphalene derivatives, may be generally incorporated into β- and γ-cyclodextrins. The host-guest complex formation between four types of anthocyanins, and α-, β-, γ-cyclodextrins were examined, and investigated specifically towards the enhancement of color stability, and the functionality of anthocyanins as food factors. Figures 6 and 7 show the stability of anthocyanins (10-4 mole/L) at pH 3 and pH 5 with β-cyclodextrin in 10mM cyclodextrin concentrations. Without the cyclodextrin, the color stability was very different from that of the β-cyclodextrin solution. At pH 5, the solution color had completely faded. At pH 3.0, four kinds of malvidin glucosides, with or without p-coumaric acid as an ester maintained solution color for a long period of time. Malvidin 3-glucoside maintained about 60% of the color. As malvidin 3-glucoside with p-coumaric acid formed a precipitate, which settled at the bottom of the solution, visible absorption was not observed. At pH 3.0, chemical equilibrium between the flavylium (colored, F) form and the pseudobase (colorless, P) form of the four anthocyanins was quickly attained after dissolution. The equilibrium can be expressed by:

$$K(\text{equilibrium constant}) = F / P$$

Proton signals from both forms, in a weakly basic solution, using 1H-NMR, were observed. Color stability, in this case, is represented by the ratio of the concentrations of the flavylium form to the pseudobase form of the anthocyanin (26). If the equilibrium shifts to the flavylium form, the color retention should be increased. Although the

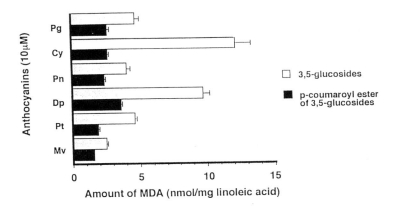

Figure 3. Comparison of antioxidative activity between deacylatedanthocyanins and acylated anthocyanins in the linoleic acid systemat pH7.4. Lipid peroxidation was induced by 400μM FeCl2 (n=4).
Pg: Pelargonidin, Cy: cyanidin, Pn: peonidin, Dp: delphinidin, Pt: petunidin, Mv: malvidin

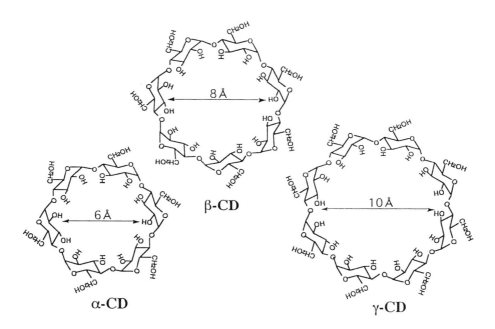

Figure 4. Chemical structure of Cyclodextrins

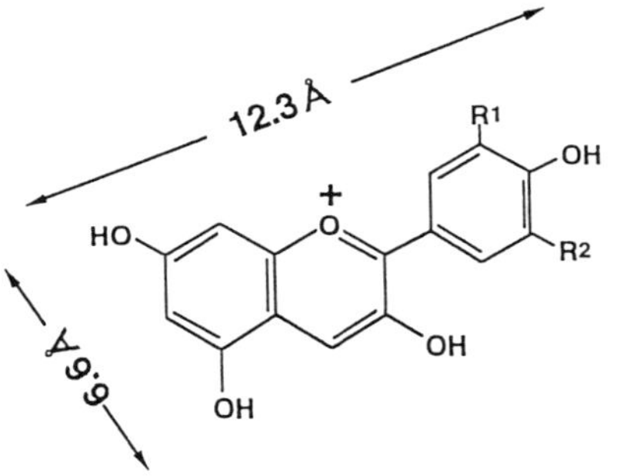

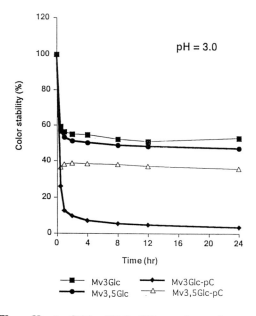

	R1	R2
Pelargonidin	H	H
Cyanidin	OH	H
Delphinidin	OH	OH
Peonidin	H	OCH3
Petunidin	OH	OCH3
Malvidin	OCH3	OCH3

Figure 5. Molecular size of anthocyanidins

Figure 6. The effect of 10mM β-CD on the color stability of malvidin type of anthocyanins. Conc =1X10^{-4}M, Temp = 30°C, pH = 3.0

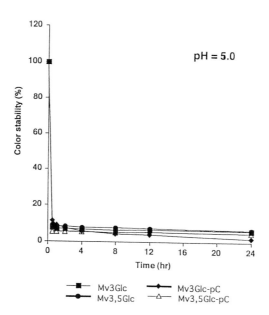

Figure 7. The effect of 10mM β-CD on the color stability of malvidin type of anthocyanins. Conc =1X10^{-4}M, Temp = 30°C, pH = 5.0

preferential inclusion of the colorless forms of the pigment into the β-cyclodextrin cavity has been reported (27), the instability of the colored forms of those anthocyanins in solution with β-cyclodextrin was not observed in this study. Thus, the equilibrium constant of the flavylium and pseudobase forms did not change significantly, despite the presence of β-cyclodextrin. However, in the buffer solutions at pH 3 and 5, γ-cyclodextrin had a slight stabilizing effect on some colored anthocyanin molecules. The stability depended on the mole ratio of γ-cyclodextrin to the pigments (1×10^{-4}M). Malvidin 3-(6- O-p-coumaroylglucoside) possessed the most stable color by adding 10mM of γ-cyclodextrin in the solution, as shown in Figs. 8 and 9. Mistry et al. and Dangles and Brouillard (15, 16) tested the inclusion effect of malvin into the γ-cyclodextrin cavity. They did not observe an improvement in the stability of malvin when bound in the γ-cyclodextrin cavity. Actually, malvidin 3,5-diglucoside showed the weakest stability in our experiments. It is possible that malvidin 3,5-diglucoside could not be included into the γ-cyclodextrin cave because of steric hindrance by bulky 5-glucoside within the cavity. However, the inclusion complex formation of malvidin 3-(6-O-p-coumaroylglucoside), and γ-cyclodextrin did cause an increase in solubility into the buffer solution. That is, solubility of the complex was higher than that of malvidin 3-(6-O-p-coumaroylglucoside) itself for the buffer solution (90% of the color was retained).

The 1H-NMR spectrum of the inclusion complex of malvidin 3-(6-O-p-coumaroylglucoside) and γ-cyclodextrin showed higher field shift of the 3-position methyne proton signal, of the glucose unit of γ-cyclodextrin, indicating an inclusion complex (28). These trends in color stability were similar in pH 5 solutions. The inclusion complex of malvidin 3-(6-O-p-coumaroylglucoside), and γ-cyclodextrin, showed color retention at about 30% at pH 5.0, even though the other anthocyanins lost about 95% of their color within 24 h of dissolving. As mentioned above, γ-cyclodextrin can incorporate two molecules of nathphalene into the cave. Therefore, the p-coumaroyl moiety of malvidin 3-(6-O-p-coumaroylglucoside) might take part as an intramolecular spacer of malvidin 3-glucoside (29). The rigid packing of the p-coumaroyl moiety and malvidin chromophore into the cave increased the color stability of the complex in solution at pH 3.0 and 5.0. From this, a tentative mechanism for the color stability of malvidin 3-(6-O-p-coumaroylglucoside) with γ-cyclodextrin can be proposed as shown in Fig. 10.

Antioxidative activity of anthocyanins in buffer solutions with cyclodextrin

Antioxidative activities of anthocyanins, with or without γ-cyclodextrin, was examined as follows (Fig. 11). Lipid peroxidation was induced by 150 mM $FeCl_2$ at 30°C for 16 h in the dark. Antioxidative activity was observed only at pH 7.4 (physiological condition). Under the physiological condition, γ-cyclodextrin (10mM) showed antioxidative activity, but this concentration of γ-cyclodextrin produced 20 nmols MDA/mg linoleic acid. After the addition of four kinds of malvidin glucosides, with or without p-coumaric acid, lipid peroxidation with the anthocyanin-γ-cyclodextrin complex was found to be suppressed. This suggests that, in the inclusion molecule, anthocyanin still acted as an antioxidative reagent. However, the antioxidative power was slightly less relative to the condition without γ-cyclodextrin.

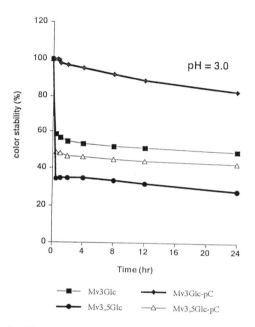

Figure 8. The effect of 10mM γ-CD on the color stability of malvidin type of anthocyanins. Conc =1X10^{-4}M, Temp = 30°C, pH = 3.0

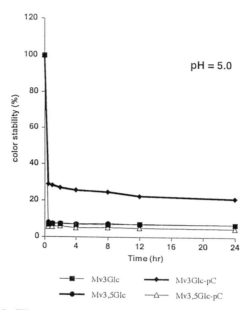

Figure 9. The effect of 10mM γ-CD on the color stability of malvidin type of anthocyanins. Conc =1X10^{-4}M, Temp = 30°C, pH = 5.0

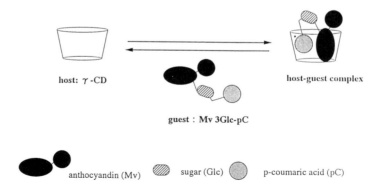

Figure. 10.   Tentative mechanism for the color stability of malvidin 3-O-(6-O-p-coumaroylglucoside) with γ-cyclodextrin.

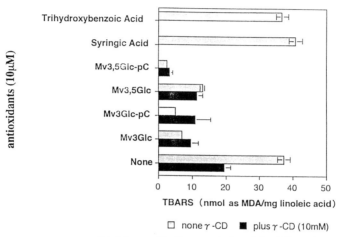

TBARS (nmol as MDA/mg linoleic acid)

□ none γ -CD   ■ plus γ -CD (10mM)

Lipid peroxidation was induced by 150 mM FeCl₂ at 30°C
and pH 7.4 for 16 hr under the dark place .

**Figure 11. The effect of γ-CD on antioxidative activity of
malvidin type of anthocyanins in the linoleic acid system.**

Conclusion and Perspective

The color stability of acylated malvidin 3-glucoside was extremely enhanced by the formation of an inclusion molecule at pHs 3.0 and 5.0. However, under physiological conditions (i.e., pH 7.4), the color stability was not as great. In contrast, antioxidative activity of anthocyanins at pH 7.4 still remained after formation of the host-guest complex with γ-cyclodextrin. For the purposes of an industrial application, the utilization of grape anthocyanin pigments at lower pHs could be investigated in future study. The shelf-life of grape color in processed foods might be increased by inclusion into a γ-cyclodextrin cavity. Furthermore, after digestion, the host-guest inclusion complex between malvidin 3-(6-O-p-coumaroylglucoside)-5-glucoside, or malvidin 3-(6-O-p-coumaroylglucoside), and γ-cyclodextrins may still show the functionality of anthocyanins as food factors for human health.

Acknowledgment

This work was partially supported by a Grant-in-Aid for Towa-Shokuhin Kenkyushinkokai, Tokyo, Japan.

Literature Cited

1. Renaud, S.; Lorgeril, M.D. Lancet, 1992; 339, 1523-1526
2. Frankel, E.N.; Kanner, J.; German, J.B.; Parks, E, Kinsella, J.E. Lancet, 1993, 341, 454-457.
3. Esterbauer, H.; Gebicki, J.; Puhl, H.; Jrgens, G. Free Radical Biol. Med., 1992; 13, 341-390.
4. Leger, A.S.ST.; Cochrane, A.L.; Moor, F. Lancet, 1979, 1017-1020
5. Seigneur, M.; Bonnet J.; Dorian, B.; Benchimol, D.; Drouillet, F.; Gouverneur, G.; Larrue, J.; Crockett, R.; Boisseau, M.R.; Ribereau-Gayon, P.; Bricaud, H. J. Appl. Cardiology, 1990, 5, 215-222
6. Timberlake, C.F.; Bridle, P. Am. J. Enol. Vitic. 1976, 27, 97-105.
7. Somer, T.C. Phytochemistry, 1971, 10, 2175-2186.
8. Fuleki, T.; Ricardo da Silva, J.M. J. Agirc. Food Chem., 1997, 45, 1156-1160
9. Duke, J. A. In Handbook of Biologically Active Phytochemicals and their Activities, 1992; CRC press
10. Middleton J. E.; Kandaswami, C. In The flavonoids: Advances in research since 1986: Harborne, J.B., Ed., Chapman & Hall; London, 1994, 619-652
11. Morazzoni, P.; Livio, S.; Scilingo, A.; Malandrino, S. Arzneim.-Forsch./Drug Res. , 1991, 41, 128-131.
12. Igarashi, K.; Takanashi, K.; Makino, M.; Yasui, T. Nippon Shokuhin Kogyo Gakkaishi, 1989, 36, 852-856.
13. Yamada, T.; Komiya, T.; Akaki, M. Agric. Biol. Chem., 1980, 44, 1411.
14. Dangles, O.; Wigand, M.C.; Cheminat, A.; Brouillard, R. Bull. Liaison Groupe Polyphenols, 1990, 15, 336-343.
15. Dagnles, O.; Brouillard, R. J. Chem. Soc., Perkin Trans. I I, 1992, 247- 257
16. Mistry, T.V.; Cai, Y.; Lilley, T.H.; Haslam, E. J.Chem.Soc., Perkin Trans. I I, 1991, 1287-1296
17. Goto, T.; Kondo, T.; Tamura, H.; Imagawa, H.; Iino, A.; Takeda, K. Tetrahedron Lett., 1982, 36, 3695-3698

18. Tamura, H.; Hayashi, Y.; Sugisawa, H.; Kondo, T. Phytochemical Anal., 1994, 5, 190-196.
19. Tamura H.; Shibamoto T. Lipids, 1991, 26, 170-173.
20. Tamura, H.; Kitta, K.; Shibamoto, T. J. Agric. Food Chem. 1991, 39, 439-442.
21. Tamura H.; Shibamoto T. J. Am. Oil Chem. Soc., 1991, 68, 941-943.
22. Tamura H.; Yamagami A. J. Agric. Food Chem., 1994; 42, 1612-1615
23. Ohkawa, H.; Ohishi, N.; Yagi, K. Anal. Biochem. 1979, 95, 351-358.
24. Burton, G.W.; Ingold, K.U. J. Am. Chem. Soc., 1981, 103, 6472-6477.
25. Ueno, A.; Takahashi, K.; Osa, T. J. Chem. Soc. Chem. Commun., 1980, 921-922
26. Brouillard, R.; Delaporte, B. J. Am. Chem. Soc. 1977, 99, 8461-8468
27. Brouillard, R.; Dagnles, O.; In The flavonoids, Advances in research since 1986, Harborne, J.B.Ed.; Chapman & Hall; London; 1994, 565-588
28. Bergeron, R.; Rowan, R. III, Bioorg. Chem., 1976, 5, 425-436
29. Ueno, A.; Moriwaki, F.; Hino, Y.; Osa, T. J. Chem. Soc. Perkin Trans II, 1985, 921- 923

Chapter 17

Inhibitory Effect of Flavonoids on Lipoxygenase-Dependent Lipid Peroxidation of Human Plasma Lipoproteins

J. Terao[1], E. L. da Silva, H. Arai, J-H. Moon, and M. K. Piskula

National Food Research Institute, Ministry of Agriculture, Forestry and Fisheries, 2-1-2 Kannondai, Tsukuba, Ibaraki 305, Japan

There is increasing evidence that oxidative modification of plasma low-density lipoprotein (LDL) leads to the formation of lipid-laden foam cells in atherosclerotic lesions. In recent years, a role of lipoxygenase in this event has attracted much attention, although the mechanism for the modification still remains uncertain. Thus, we evaluated the inhibitory effect of several dietary antioxidants, including flavanoid aglycone, and glycosides, on the lipoxygenase-dependent lipid peroxidation of human plasma LDL. 15-Lipoxygenases (15-LOX) from soybean and rabbit reticulocytes were used as the enzymes. LDL oxidation was monitored by the measurement of cholesteryl ester hydroperoxides (CE-OOH), using reverse-phase HPLC. (-)-Epicatechin, quercetin, and quercetin monoglucosides (quercetin-3-O-β-glucopyranoside, quercetin 4'-O-β-glucopyranoside, quercetin-7-O-β-glucopyranoside) were found to inhibit the accumulation of CE-OOH, whereas no inhibition was observed with 5-fold α-tocopherol-enriched LDL. It is, therefore, implied that dietary flavonoids, such as catechins and quercetin, can exert a role in decreasing cardiovascular diseases by inhibiting lipoxygenase-induced LDL oxidation.

It is generally accepted that oxidatively modified, low-density lipoprotein (LDL) plays an important role in the early stages of atherogenesis (1-3). Although the exact mechanism by which oxidized LDL is formed *in vivo* is still unknown, LDL can be modified oxidatively by 15-lipoxygenases (15-LOX) in *in vitro* systems (4-7). High levels of lipoperoxides were found in LDL, and incubated with fibroblasts which overexpress 15-LOX (8,9). Therefore, the 15-LOX reaction seems to participate in the initial process of LDL oxidation as shown if Fig. 1. Several clinical trial studies have indicated that dietary antioxidants help in preventing the progression of coronary artery disease (review, see ref. 10). These studies are focusing on the introduction into the LDL particles of chain-breaking antioxidants, such as α-tocopherol and the carotenoids, which decrease the extent of LDL oxidation. The use of 15-LOX inhibitors from dietary sources might be an interesting mode of prevention of atherosclerosis if 15-LOX plays an essential part in oxidative modification of LDL *in vivo*. Here, we selected (-)-epicatechin, quercetin, and quercetin monoglucosides as typical dietary flavonoids (Figure 2), which are recognized as lipoxygenase inhibitors (11,12). Their inhibitory effects on 15-LOX-dependent LDL oxidation was evaluated by measuring the accumulation of cholesteryl ester hydroperoxides (CE-OOH). The results suggest that on a molar basis, these flavonoids are stronger inhibitors than α-tocopherol in the 15-LOX-induced oxidative modification of LDL.

Effect of Lipophilic Antioxidants on Soybean Lipoxygenase-Dependent Peroxidation of Phospholipid-Bile Salt Micelles

We first investigated the inhibition of some typical antioxidants in the enzymatic lipid peroxidation of bile-salt micelles of phosphatidylcholine (PC), using soybean lipoxygenase (13). Soybean enzyme can oxidize micellar phospholipids directly in the presence of bile salt (14,15). The inhibition ratio of each antioxidant was calculated from the phosphatidylcholine hydroperoxide (PC-OOH) concentration after 60 min. incubation with, and without, antioxidants. The 50% inhibition concentrations (IC50) were obtained, as shown, in Table 1. The IC50 of quercetin (90 μM), is much lower than α-tocopherol (220 μM). This indicates that quercetin possesses a higher inhibitory effect than α-tocopherol in the 15-LOX-dependent lipid peroxidation of esterefied lipids, in biological systems. β-Carotene showed no inhibitory effect, even at the concentration of 1 mM.

Effect of Antioxidants on Soybean Lipoxygenase-Dependent Peroxidation of Human LDL.

Human LDL was isolated from fresh human plasma by discontinuous density-gradient ultracentrifugation, according to the method of Kleinveld (16). Its suspension in PBS buffer (0.2 mg protein/ml) was oxidized by the addition of 10,000 U/ml of soybean lipoxygenase (Typel-B, 206,000 units/mg protein from Sigma Chemical Co.). LDL oxidation was carried out at 20°C, and the cholesteryl ester hydroperoxides (CE-OOH) was measured by reverse phase HPLC as the index of LDL oxidation levels (Figure 3). There was a progressive increase in the CE-OOH level without a detectable lag phase. In addition, 1 μM ascorbic acid was an effective antioxidant only in the first 3 hours. The LDL fraction used in the experiment, contained endogenous α-tocopherol, at the level of 11.4 n mol/mg protein.

We prepared α-tocopherol-enriched LDL by the addition of D/L-α-tocopherol to plasma. LDL particles were isolated after incubation for 3 hours. This LDL contained 5-fold enriched α-tocopherol (50.2 nmol/mg protein), however, no inhibitory effects were identified by the enrichment treatment. In contrast, (-)-

174

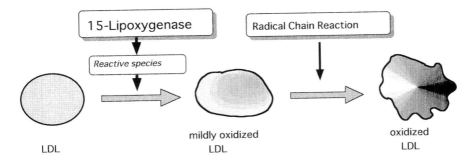

Fig. 1 Structure of Flavonoids Used in This Study

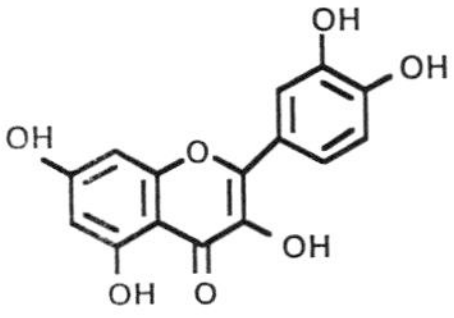

Quercetin 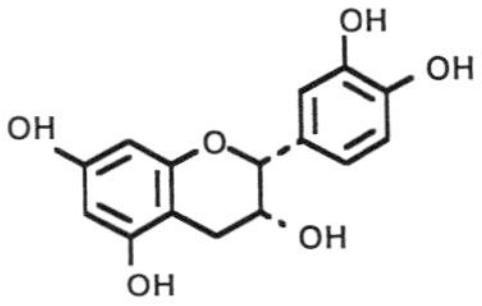   (-)-Epicatechin 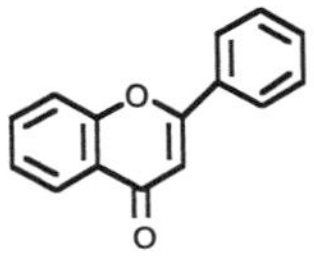   Flavone

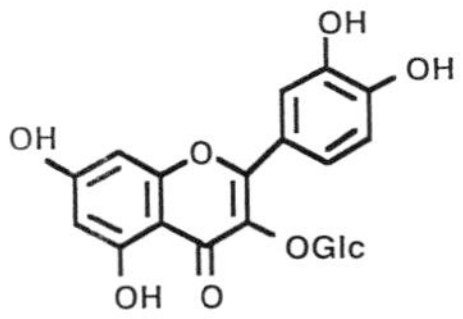

Q3G 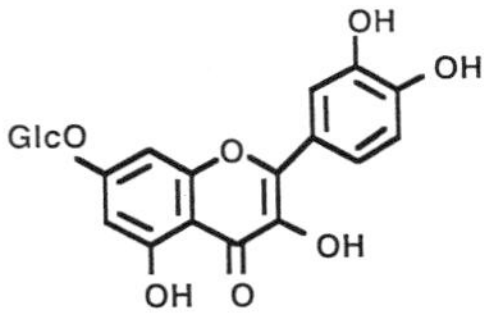   Q7G 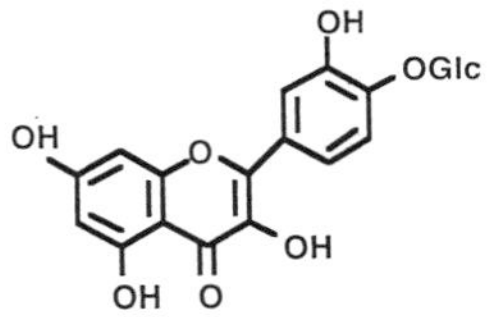   Q4'G

Fig. 2 Possible Pathway for Oxidative Modification of Human Low-Density Lipoprotein (LDL) Leading to Athersclerosis.

Table I. The 50% Inhibition Concentrations (IC50) of Quercetin and Quercetin Monoglucosides on Mammalian 15-LOX-Induced Oxidation of Cholesteryl Esters in Human LDL

Antioxidants	IC50 (μM)
Quercetin	0.35
Q3G	0.47
Q7G	0.47
Q4'G	1.2

Human LDL (0.4 μM) was oxidized with 15-LOX (1 μM) at 20°C, for 3 hours in the presence or absence of quercetin or quercetin glucosides.

epicatechin and quercetin were found to be strong LOX inhibitors. Thus, it is indicated that (-)-epicatechin and quercetin act as effective inhibitors when LDL was subject to the lipoxygenase reaction.

Effect of Antioxidants on Mammalian 15-LOX-Dependent Lipid Peroxidation of Human LDL.

Mammalian 15-LOX has been purified from various tissues, and its molecular properties have been studied (17). Mammalian 15-LOX is capable of oxygenating esterified fatty acids, not only in micelles, but also in biomembranes and plasma lipoproteins. It is suggested to participated in the oxidative modification of LDL. We applied LOX from rabbit reticulocytess as mammalian 15-LOX, and its reaction with LDL was used as the *in vitro* model system for the estimation of the antioxidant activity of flavonoids in the LOX-induced oxidative modification of LDL (Figure 4). Similar to soybean LOX, CE-OOH accumulated linearly by the reaction of reticulocyte 15-LOX with human LDL, and (-)-epicatechin and quercetin, at 1 μM, and inhibited the accumulation effectively. α-Tocopherol-enriched LDL did not inhibit the CE-OOH accumulation, and ascorbic acid only inhibited the first stage of 3 hours, at the level of 1 μM. Thus, flavonoids are superior to α-tocopherol and ascorbic acid in the 15-LOX-dependent oxidation of LDL. However, flavone at the same concentration was completely ineffective. Indeed, flavone contains a flavonoid nucleus, but does not have any additional phenolic groups. It is therefore likely that the substitution of the flavone nucleus with hydroxyl groups is essential to the ability of flavonoids to inhibit the 15-LOX catalyzed LDL oxidation.

Effect of Quercetin and Quercetin Monoglucosides on Mammalian 15-LOX-Dependent Lipid Peroxidation of Human LDL

A number of flavonoids are commonly found in foods in the form of glycosides. For example, quercetin is mainly present in the forms of 3,4'-di-*O*-β-glucosides, and 4'-*O*-β-glucoside in onion (18). Glucosidase from human plasma, and intestinal bacteria, are likely to hydrolyze the glucose moiety, resulting in a flavonoid aglycone (19). However, a recent study has shown the presence of flavonoids as glycosides in human plasma (20). In order to know the difference between a flavonoid aglycone, and its glycoside, in the inhibition of 15-LOX-induced LDL oxidation, we selected quercetin and quercetin glucosides (quercetin-3-*O*-β-glucopyranoside; Q3G, quercetin 4'-*O*-β-glucopyranoside; Q4G, and quercetin-7-*O*-β-glucopyranoside; Q7G).

Table 1 shows the IC 50 (the 50% inhibition concentration) of these aglycones and glucosides. Interestingly, binding of the glucose group at the 3- and 7- positions influenced the inhibitory effect very little, indicating that hydroxyl groups at 3- and the 7- positions, at least partly, do not take an essential part in the inhibition mechanism. On the other hand, IC 50 obtained from Q4'G was lower than those obtained from the other glucosides and aglycone. The hydroxyl group at the B ring seems to, to some extent, contribute to the inhibition of 15-LOX-induced oxidation.

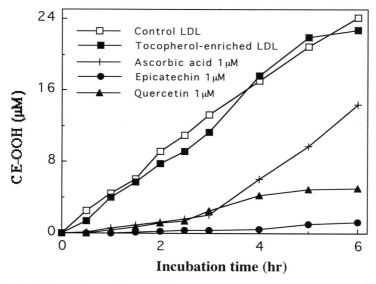

Fig. 3 Effect of antioxidants on LDL oxidation indiced by soybean 15-LOX.

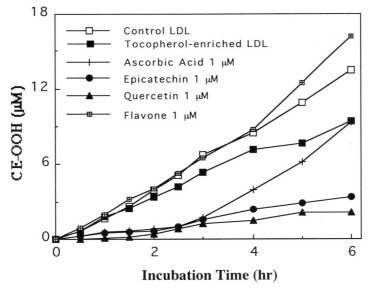

Fig. 4. Effect of antioxidants on LDL oxidation induced by rabbit reticulocyte 15-LOX.

Conclusion

In 15-LOX-induced LDL oxidation, flavonoids containing a polyhydroxyl group can act as strong inhibitors. They may, therefore, have a role in the prevention of atherosclerosis if 15-LOX participates in the process of this degenerative disease. Endogenous α-tocopherol in LDL may not work as an effective inhibitor of this process. Quercetin glucosides, as well as aglycone, can inhibit this LOX-induced oxidation. These flavonoids may be promising compounds for the prevention of atherosclerosis.

Literature Cited

1. Ross, R. *Nature*, **1993**, 356, 801-809.
2. Holvoet, P.; Collen, D. *FASEB J.*, **1994**, 8, 1279-1284.
3. Steinberg, D.; Parthasarathy, S.; Carew, T.E.; Khoo, J.C.; Witztum, J.L *N. Engl. J. Med.* **1989**, 320, 915-924.
4. Parthasarathy, S.; Wieland, E.; Steinberg, D. *Proc. Natl. Acad. Sci. USA,* **1989**, 86, 1046-1050.
5. Belkner, J.; Wiesner, R.; Rathman, J.; Barnett, J.; Sigal, E.; Kuhn, H. *Eur. J. Biochem.* **1993**, 213, 261-261.
6. Kuhn, H.; Belkner, J.; Suzuki, H.; Yamamoto, S. *J. Lipid Res.* **1994**, 35, 1749-1759.
7. Upston, J.M.; Neuzil, J.; Stocker, R. *J. Lipid Res.* **1996**, 37,2650-2661.
8. Benz, D.J.; Mol., M.; Ezaki, M.; Mori-ito, N.; Zelan, I.; Miyanohara, A.; Friedmann, T.; Parthasarathy, S.; Steinberg, D.; Witzum, J.L. *J. Biol. Chem.* **1995**, 270, 5191-5197.
9. Ezaki, M.; Witztum, J.L.; Steingerg, D. *J. Lipid Res.* **1995**, 36, 1996-2004.
10. Duel, P.B. *J. Nutr.* **1996**, 126, 1067S-1071S.
11. Yoshimoto, T.; Furukawa, M.; Yamamoto, S.; Horie, T.; Watanabe-Kohno, S. *Biochem. Biophys. Res. Commun.* **1983**, 116, 612-618.
12. Laughton, M.J.; Evans, P.A.; Moroney, M.A.; Hoult, J.R.S.; Halliwell, B. *Biochem. Pharmacol.* **1991**, 42, 1673-1681.
13. Arai, H.; Nagao, A.; Terao, J.; Suzuki, T.; Takama, K. *Lipids* **1995**, 30, 135-140.
14. Eskola, J.; Laakso, S. *Biochim. Biophys. Acta* **1983**, 751, 305-311.
15. Brash, A.R.; Ingram, C.D.; Harris, T.M. *Biochemistry* **1987**, 26, 5465-5471.
16. Kleinveld, H.A.; Hak-Lemmers, H.L.; Stalenhoef, A.F.H.; Demacker, N.M. *Clin. Chem.* **1992**, 38, 2066-2072.
17. Yamamoto, S. *Biochim. Biophys. Acta* **1992**, 1128, 117-131.
18. Tsushida, T.; Suzuki, M. *Nippon Shokuhinkogaku Kaishi* **1996**, 43, 642-649.
19. Tamara, G.; Gold, C.; Ferro-Luzzi, A.; Ames, B.N. *Proc. Natl. Acad. Sci. USA,* **1990**, 77, 4961-4965.
20. Paganga, G.; Rice-Evans, C.A. *FEBS Lett.* **1997**, 401, 78-82.

Chapter 18

Comparison on the Nitrite Scavenging Activity and Antimutagen Formation Between Insoluble and Soluble Dietary Fibers

K. Kangsadalampai

Division of Food and Nutrition Toxicology, Institute of Nutrition, Mahidol University, Puttamonthon 4 Road, Salaya, Nakhon Pathom 73170, Thailand

Nine insoluble fibers prepared from fruits and vegetables and eleven soluble fibers were evaluated for their nitrite scavenging activity and the antimutagen formation of aminopyrene-nitrite (AP-nitrite) model in the Ames test. Most of the insoluble fibers were good nitrite scavengers and inhibitors of mutagen which arose from the reaction of the model (pH 3). However, it was found that most of the soluble fibers were not good nitrite scavengers. The presence of some fibers (7 out of 11) in the reaction mixture of the AP-nitrite model increased the revertant number of *Salmonella typhimurium* TA98 and TA100 induced by the reaction products. Subsequent experiments showed that the binding of aminopyrene to some soluble fibers was reversible and might be the possible mechanism of mutagenic enhancement.

Dietary fiber is believed to decrease the incidence of colorectal cancer; but not all types of fiber are equally protective. Two studies questioned the philosophy that a high fiber diet will do no harm, even it does not prevent colon cancer. An Australian study showed a positive correlation between cereal fiber intake and colon cancer (*1*) and a British study showed increased mortality from gastric cancer in vegetarians (*2*). Studies on the antimutagen formation of fiber (especially fiber from fruits and vegetables) are scanty. Moller et al. (*3*) noted that wheat bran acted as a nitrite scarvenger under gastric-like conditions. Their work indicated that nitrite scavenging may be a factor in the possible role of bran protecting against gastric cancer development. It was observed in an *in vitro* experiment that hydrophobic carcinogens adsorbed to the insoluble plant cell wall component, but commercial soluble fiber currently used as emulsifiers and stabilizers in the food industry namely; κ-carrageenan, γ-galactomannan, $(1\rightarrow3,1\rightarrow4)$-β-D-glucan, gum arabic, pectin, polygalacturonic acid and sodium carboxymethylcellulose, maintained 1,8 dinitropyrene in aqueous solution and decreased its adsorption to α-cellulose (*4*). This phenomenon suggested the possible mechanisms by which soluble fibers might enhancethe development of cancer. It was thus worth examining the role of dietary fibers as nitrite scavengers and their antimutagen formation.

Materials and Methods

Chemicals. Bacto agar was the product of Difco Laboratories, (Detroit, Michigan, U.S.A.). Oxoid nutrient broth No. 2 was purchased from Oxoid Ltd., (Basingstoke, Hampshire, England). Aminopyrene was furnished by Aldrich (St. Louis Missuri, U.S.A.). Sodium thiocyanate (NaSCN), bovine serum albulmin (BSA), d-biotin, sodium dihydrogen phosphate and α-cellulose were purchased from Sigma Chemical Co. (St. Louis, Missouri, U.S.A.). Hydrogen bromide, sodium nitrite and N-(1-napthyl)ethylenediamine dihydrochloride were obtained from BDH Chemicals Ltd. (Poole, England). Other chemicals were of laboratory grade.

Plant Fibers. They were prepared from cabbage, Chinese kale, cucumber, waxgourd, ivygourd, papaya, unpolished rice, rice bran, gauva, pineapple core and pomelo peels. Each sample was disrupted and blended with water; the homogenate was then frozen overnight. After thawing, it was washed repeatedly with water to remove water-soluble materials including protein and water soluble vitamins (5). The fiber was extracted with a mixture of hexane and acetone (3:2) to remove lipid soluble compounds. Finally the resultant powder was ground and passed through a 20-mesh screen.

Soluble Fibers. They were obtained from various sources. Agar was the product of Difco Laboratories (Michigan, U.S.A.). Carboxymethylcellulose (CMC) was prepared by D.S.K. Internationals (Tokyo, Japan). Carrageenan was furnished by Systems Bioindustries (Baupte, France). Gum arabic and guar gum were supplied by Rickard Gum (England) and Jainson (Jodhiur, India) respectively. Locust bean gum was bought from C.E. Roepera GmbH and Co. (Hamburg, Germany). Three methylcellulose namely; MC 25 cps (centipoise), MC 1500 cps (centipoise) and MC 4000 cps (centipoise) were the products of Dow Chemical (Michigan, U.S.A.). Pectin was purchased from Hercules, Inc. (New York, U.S.A.). Sodium alginate and xanthan gum were the products of Went Cheme (Germany) and C.N.I. (Paris, France).

Nitrite Scavenging Capability (NSC). Fiber (20, 40 or 100 mg) was added to 5 ml of simulated gastric mixture containing BSA 300 mg/l and imitated saliva (consisting of 0.3 mM NaSCN and 2 g/l NaCl) as suggested by Moller et al. (3). The mixture was refrigerated overnight. Before incubation, 5 ml of 80 mM sodium nitrite was added to the reaction mixture and pH level was adjusted to 1.5-2.0 with 0.5 ml of 2N HCl. Incubation through shaking was conducted at 37° C for 2 h. The reaction tube was placed in an ice bath for 10 min before the reaction mixture was finally filtered through filter paper. The procedure by Takeda and Kanaya (6) was used to determine the nitrite content. NSC was the slope obtained from the linear regression between calculated amount (in mg) of fiber and the amount (in μg) of nitrite that disappeared from the reaction mixture.

180

Antimutagen Formation of Fibers. Aminopyrene interacting with sodium nitrite in an acidic medium (pH 3) was used as a standard mutagenic model (7). Fiber sample (10, 20, 40, 80 or 120 mg) was added to 2 ml of simulated gastric condition mixture containing BSA 300 mg per litre and imitated saliva (3). The mixture was refrigerated overnight. Before incubation, 300 µl of 0.004 M sodium nitrite, 300 µl of 0.004 M aminopyrene and 1400 µl of 0.02 N hydrochloric acid (sufficient acid to acidify the reaction mixture to a pH level of 3.0-3.5) were added. Thorough shaking during incubation was carried out at 37° C for 4 h. The mixture was then placed in an ice bath for 10 min. The reaction tubes were centrifuged and the supernatant was collected and filtered through sterile 0.2 micron filter paper. Filtrate (0.015 ml or 0.025 ml) of the reaction mixture was mixed with 0.5 ml of 0.5 M NaPO$_4$-KCL buffer pH 7.4 and was subjected to preincubation method of the Ames mutagenicity test using *Salmonella typhimurium* strain TA 98 or TA 100 modified by Yahagi et al. (8) without the addition of activating system. The fiber incorporated in the reaction mixture tube was said to have antimutagen formation activity when the *Salmomella* revertants per plate were lower than those of the standard nitrite treated aminopyrene plate and demonstrated a negative dose-response relationship.

Aminopyrene Binding Strength of Different Fibers. The experiment was designed to elucidate a possible mechanism to explain why 9 out of 12 fibers could increase the formation of mutagen. It was conducted simultaneously in two separated tubes. In each tube, fiber sample (15 mg per tube) was autoclaved at 121 °C for 15 min, brought to room temperature, and added with 2.0 ml of sterile water and the tube was kept refrigerated overnight. The solutions of 10 µl of aminopyrene (0.075 mg per ml) and 740 µl of 0.2 N hydrochloric acid were added to the tube. Each reaction tube was first shaken in the waterbath for 1 h. It was then centrifuged at 3,500 rpm, for 20 min. One ml of the supernatant of the first tube (Portion A) was transferred to the tube containing 90 µl of 2 M sodium nitrite and refrigerated. To the remaining mixture in the first tube, 1 ml of sterile distilled water (pH 3.0) was added and further incubation was carried out for another 1 h; then, the mixture was centrifuged and 1 ml of the supernatant was taken as Portion B. This portion was also added 90 µl of 2 M sodium nitrite. The remaining second tube was incubated in a shaking water bath (37°C) for 2 h, centrifuged and 1 ml of the supernatant (Portion C) was added with 90 µl of 2 M sodium nitrite. The reaction mixtures of Portions A, B and C were shaken for another 2 h in the water bath at 37 °C before they were mixed with 90 µl of 2 M ammonium sulfamate in the ice bath for 10 min. Mutagenicity of each portion was determined as described above.

Results and Discussion

Nitrite Scavenging Capability. The NSC results of various fiber types are shown in Table I. The NSC of ivygourd fiber is noted as the best. Indications note that fiber from various fruits and vegetables show different NSC. NSC (ng nitrite/mg fiber) ranged from 135.8 for cooked pomelo peels to 252.4 of ivygourd. A widely used food additive α-cellulose; however, had very low capability (1.6 ng nitrite/mg fiber). On the other hand, it was found that no soluble fibers changed the concentration of nitrite under the studied condition. It is important to note that the studied plant fibers are insoluble fibers according to the method of preparation which could remove all water soluble substances. From the present experiment, it appears that only insoluble fibers are beneficial as nitrite scavengers.

The study results have demonstrated that plant fiber are nitrite scavengers under conditions similar to those prevailing in a normal human stomach. Dence (9) documented that wood lignin could react with nitrite under markedly acidic conditions. However, most of the fiber samples in this study were prepared from unlignified wall of dicotyledonous plants (except for rice bran and unpolished rice) which were poor lignin sources. The only possible lignin source in this study might be the pineapple core which was, nevertheless, not a good nitrite scavenger. Further elucidation on the components of the fiber samples which are nitrite scavengers is required.

Antimutagen Formation of Fiber It is shown that most fiber extracted from various plant sources can bind nitrite. The inverse dose-response relationships between the amount of fiber and the mutagenicity of the nitrite treated aminopyrene are shown in Table II. It is noted that plant fibers have significant antiformation of mutagens on both tester strains of Ames test. In addition, it is not surprising to find that α-cellulose has no antiformation of mutagen. It is noted that pineapple core fiber, which possibly contained high lignified cell walls, trapped nitrite and inhibited the mutagenicity of the studied model. However, its NSC and antiformation of mutagen were not better than those of the others. Table III shows that three fibers namely; carrageenan, sodium alginate, and xanthan gum exhibit their inhibitory effects against the mutagenicity of the model. However, the fibers namely; agar, CMC, guar gum, gum arabic, locust bean gum, MC 25, MC 1500, MC 4000, and pectin increase the mutagenicity of nitrite treated aminopyrene on *Salmonella typhimurium* TA98 and TA100.

Rice bran was a good source of nitrite scavenging fiber. Its extracted fiber was the second best in reducing the mutagenicity of nitrite treated aminopyrene. Moorman et al. (10) worked on the ion exchange capacity and ability to bind mutagens such as benzo[a]pyrene and 2-aminoanthracene (2-AA) and mutagens extracted from fried ground beef by soft white wheat bran, hard red wheat bran, and corn bran. They found that wheat bran had the highest ion exchange capacity and were effective in binding the 2-AA. Benzo[a]pyrene and the fried ground beef mutagens were not effectively bound by fibers which were demonstrated by the failure of reducing the mutagenicity by passing the hamburger mutagen through the

Table I. Nitrite scavenging capability of different fibers.
Data were the slope of the linear regression between the amount of
fiber (in milligram) per tube and the disappearance of
nitrite ion (in nanogram) from the reaction mixture.
Data were reported as means and standard deviations (n = 4)

Fiber	NSC (ng nitrite per mg fiber)
cabbage	213.0 ± 22.1
Chinese kale	237.6 ± 31.2
pomelo peel	91.1 ± 18.9
cucumber	218.9 ± 25.9
waxgourd	204.7 ± 24.6
ivygourd	252.4 ± 28.4
papaya	210.1 ± 36.5
unpolished rice (boiled)	72.4 ± 15.7
rice bran	262.7 ± 17.9
pineapple core	142.2 ± 24.0
α-cellulose	1.6 ± 0.8
agar	trace*
pectin	trace
carrageenan	trace
sodium alginate	trace
guar gum	trace
locust bean gum	trace
xanthan gum	trace
gum arabic	trace
carboxymethylcellulose	trace
methylcellulose 25, 1500 and 4000 cps**	trace

* less than 1 ng nitrite per mg fiber
** cps = centipoise

Table II. Antimutagen formation of different plant fibers on 2,000 μl reaction mixture containing 300 μl of 0.004 M sodium nitrite, 300 μl of 0.004 M aminopyrene, and 1400 μl 0.02N hydrochloric acid. The fiber has antimutagen formation activity if it reduced the *Salmonella* revertants per plate with negative dose response relationship. Revertant colonies were reported as means and standard deviations in parenthesis (n = 4).

fiber	TA98			TA100		
	mg fiber per plate	revertants per plate	% reduction or stimulation	mg fiber per plate	revertants per plate	% reduction or stimulation
waxgourd	0	688(57)	0	0	697(35)	0
	38	781(11)	+13.52	63	680(12)	-2.44
	75	741(28)	+7.70	125	639(52)	-833
	150	612(57)	-11.05	250	633(28)	-9.19
	300	125(25)	-81.84	500	560(20)	-19.66
	450	18(4)	-97.39	750	270(12)	-61.27
ivygourd	0	648(54)	0	0	724(25)	0
	38	376(26)	-41.98	63	758(45)	+4.69
	75	86(13)	-86.73	125	739(14)	+2.07
	150	26(3)	-95.99	250	200(22)	-72.38
	300	20(7)	-96.92	500	184(16)	-74.59
	450	24(9)	-96.30	750	212(11)	-70.72
cabbage	0	664(44)	0	0	688(42)	0
	38	528(18)	-20.49	63	582(96)	-15.41
	75	422(23)	-36.45	125	522(46)	-24.13
	150	265(39)	-60.10	250	573(27)	-16.72
	300	32(6)	-95.19	500	474(29)	-31.11
	450	20(5)	-96.99	750	214(16)	-68.90
Chinese kale	0	750(29)	0	0	697(38)	0
	38	616(17)	-17.87	63	608(13)	-12.77
	75	434(25)	-42.14	125	589(16)	-15.50
	150	206(29)	-72.54	250	561(44)	-19.52
	300	14(2)	-98.14	500	158(18)	-77.34
	450	16(6)	-97.87	750	156(13)	-77.62
pineapple core	0	750(29)	0	0	697(38)	0
	38	606(8)	-19.20	63	539(17)	-22.67
	75	559(28)	-25.47	125	650(25)	-6.75
	150	203(21)	-72.94	250	640(22)	-8.18
	300	97(12)	-87.07	500	510(19)	-26.83
	450	40(18)	-94.67	750	330(24)	-52.66

Continued on next page.

Table II. *Continued.*

fiber	mg fiber per plate	TA98 revertants per plate	% reduction or stimulation	mg fiber per plate	TA100 revertants per plate	% reduction or stimulation
papaya	0	726(59)	0	0	688(48)	0
	38	550(38)	-24.25	63	612(36)	-11.05
	75	470(11)	-35.27	125	666(47)	-3.20
	150	134(30)	-81.55	250	528(29)	-23.26
	300	17(12)	-97.66	500	249(16)	-63.81
	450	16(8)	-97.80	750	154(18)	-77.62
pomelo peel	0	772(25)	0	0	724(32)	0
	38	740(11)	-4.15	63	710(12)	-1.94
	75	644(28)	-16.59	125	686(32)	-5.25
	150	488(11)	-36.79	250	664(14)	-8.29
	300	209(36)	-72.93	500	468(24)	-35.36
	450	92(16)	-88.09	750	240(36)	-66.86
cucumber	0	879(52)	0	0	814(71)	0
	38	624(36)	-29.02	63	790(52)	-2.95
	75	514(12)	-41.53	125	736(44)	-9.59
	150	127(25)	-85.56	250	540(16)	-33.67
	300	18(4)	-97.96	500	106(26)	-86.98
	450	21(7)	-97.62	750	85(7)	-89.56
rice bran	0	879(52)	0	0	814(71)	0
	38	692(45)	-21.28	63	791(37)	-2.83
	75	280(23)	-68.15	125	652(24)	-19.91
	150	20(12)	-97.73	250	116(18)	-85.75
	300	18(7)	-97.96	500	110(18)	-86.49
	450	14(8)	-98.41	750	91(12)	-88.83
unpolished rice (boiled)	0	664(44)	0	0	688(42)	0
	38	566(15)	-14.76	63	658(23)	-4.37
	75	537(18)	-19.13	125	640(40)	-6.98
	150	424(26)	-36.15	250	625(44)	-9.16
	300	400(29)	-39.76	500	604(37)	-12.21
	450	144(18)	-78.32	750	572(27)	-16.87
α-cellulose	0	879(52)	0	0	814(48)	0
	38	680(25)	-22.64	63	878(28)	+7.86
	75	624(21)	-29.02	125	828(35)	+1.71
	150	686(28)	-21.96	250	839(24)	+3.07
	300	762(23)	-13.32	500	806(36)	-0.99

Table III. Antimutagen formation of different soluble fibers on 2,000 μl reaction mixture containing 300 μl of 0.004 M sodium nitrite, 300 μl of 0.004 M aminopyrene, and 1400 μl 0.02N hydrochloric acid. The fiber has antimutagen formation activity if it reduced the *Salmonella* revertants per plate with negative dose response relationship. Revertant colonies were reported as means and standard deviations in parenthesis (n = 4).

fiber	TA98			TA100		
	mg fiber per plate	revertants per plate	% reduction or stimulation	mg fiber per plate	revertants per plate	% reduction or stimulation
agar	0	1176(77)	0	0	472(38)	0
	67	1400(70)	+19.04	167	564(22)	+19.49
	133	1858(46)	+57.99	333	598(13)	+16.69
	267	1595(78)	+35.62	668	592(13)	+25.42
	400	1125(28)	-4.34	1000	582(83)	+23.30
pectin	0	1271(94)	0	0	730(33)	0
	67	2672(102)	+110.22	167	1175(106)	+37.87
	133	2410(212)	+89.61	333	1620(85)	+37.87
	267	2900(98)	+128.16	668	1455(131)	+23.82
	400	3015(56)	+137.21	1000	1515(49)	+28.93
carrageenan	0	1008(94)	0	0	730(33)	0
	67	380(92)	-62.31	167	593(76)	-18.77
	133	399(93)	-60.42	333	698(46)	-4.39
	267	420(67)	-58.34	668	682(59)	-6.58
	400	350(42)	-65.28	1000	670(29)	-8.22
sodium alginate	0	788(56)	0	0	444(28)	0
	67	373(53)	-52.67	167	218(39)	-50.91
	133	226(57)	-71.32	333	224(20)	-49.55
	267	189(29)	-76.02	668	206(18)	-53.61
	400	278(20)	-64.73	1000	248(23)	-44.15
guar gum	0	1035(31)	0	0	576(35)	0
	67	1983(75)	+91.59	167	931(21)	+61.63
	133	1660(28)	+60.38	333	942(69)	+63.54
	267	778(30)	-24.84	668	492(41)	+14.59
	400	658(16)	-36.43	1000	556(21)	-3.48
locust bean gum	0	1271(94)	0	0	730(33)	0
	67	1318(154)	+3.69	167	712(54)	-1.24
	133	1182(88)	-7.01	333	802(31)	+9.86
	267	1494(111)	+17.54	668	848(65)	+16.16
	400	1722(126)	+35.48	1000	955(35)	+30.82

Continued on next page.

Table III. Continued.

fiber	mg fiber per plate	TA98 revertants per plate	% reduction or stimulation	mg fiber per plate	TA100 revertants per plate	% reduction or stimulation
xanthan	0	1008(94)	0	0	730(33)	0
gum	67	1012(69)	+0.39	167	611(26)	-16.31
	133	992(95)	-1.59	333	552(65)	-24.39
	267	726(83)	-27.98	668	490(31)	-32.88
	400	620(55)	-38.50	1000	340(21)	-53.43
gum	0	1035(32)	0	0	576(35)	0
arabic	67	1242(25)	+20.00	167	742(39)	+28.81
	133	1415(35)	+36.71	333	966(108)	+67.70
	267	1850(42)	+78.74	668	1123(24)	+94.96
	400	1412(90)	+36.42	1000	1288(44)	+123.61
carboxyl	0	780(90)	0	0	433(15)	0
methyl	67	933(24)	+19.61	167	472(11)	+9.00
cellulose	133	894(20)	+14.61	333	576(25)	+33.02
	267	1840(28)	+135.89	668	561(36)	+29.56
	400	1723(37)	+120.89	1000	617(23)	+42.49
methyl	0	780(90)	0	0	433(21)	0
cellulose	67	896(34)	+14.87	167	464(28)	+7.15
25 cps	133	918(101)	+17.69	333	454(36)	+4.84
	267	1020(91)	+30.76	668	690(28)	+59.35
	400	1640(64)	+110.25	1000	668(74)	+54.27
methyl	0	1176(77)	0	0	472(88)	0
cellulose	67	1565(108)	+33.07	167	799(30)	+69.27
1500 cps	133	1964(107)	+67.00	333	852(33)	+80.50
	267	1656(73)	+40.81	668	804(26)	+70.33
	400	1960(85)	+66.66	1000	818(11)	+73.30
methyl	0	1176(77)	0	0	472(88)	0
cellulose	67	1325(35)	+12.67	167	818(23)	+73.30
4000 cps	133	1839(100)	+56.37	333	876(13)	+85.59
	267	1990(42)	+69.21	668	680(62)	+44.06
	400	1703(137)	+44.81	1000	618(25)	+30.93

fiber coloumns. Thus, further study on the inhibition of mutagens of fried food by rice bran fiber is still required.

Aminopyrene Binding Strength of Soluble Fibers. Since nitrite scavenging capability was not a characteristic of the soluble fibers in this study, the binding to aminopyrene was of interest to be explored. Fiber-mutagen binding has been suggested as one of the mechanisms of action which may prevent mutagenesis. The results of the present study enable us to categorize the fibers into two groups (Table IV). The first group contains agar, carboxymethylcellulose, guar gum, gum arabic, locust bean gum, and pectin. These results were in accordance with the experimental design suggesting that aminopyrene bound to the fibers at a certain amount and left the unbound portion reacted with sodium nitrite and became mutagenic. When a portion of 1 ml of the mixture was removed and the mixture was made up to the original volume, the one ml portion B was found mutagenic nearly the same magnitude as of portion A. It is therefore suggested that the binding is reversible due to the concentration of aminopyrene in the environment. The second group of fiber consisting of MC 25 cps, MC 1500 cps and MC 4000 cps has a different characteristic. The mutagenicity of portion C of each fiber is not different from that of portion A. On the other hand, the mutagenicity of portion B does not recover after the addition of water suggesting that aminopyrene firmly binds to such fibers.

The results of the soluble fiber in the first group indicate a possible mechanism by which soluble fibers may enhance mutagenesis of the aminopyrene-nitrite model. The fibers may maintain aminopyrene (a hydrophobic mutagen precursor) in some unknown specific manner and presumably release it to the solution to interact with nitrite continuously. This phenomenon may occurr due to the concentration gradient between the bound aminopyrene portion and the free aminopyrene in the medium. It is obvious that the products of nitrosation of compounds other than amines in gastric-like condition are composed of some direct mutagens (*11*). Direct mutagens are generally less stable than indirect mutagens in the aqueous environment. Agents such as nitrosamides, nitrosoureas and ethylenimines are chemically stable only in the anhydrous state (*12*). Since the incubation time was 4 h, the total direct mutagen which occurred in the reaction mixture of nitrite and aminopyrene (interaction occurred for all aminopyrene at the same time) may have a lesser chance (due to its unstability) to interact with the DNA of the tester strains than that of the reaction mixture containing fiber which occurred continuously due to the released aminopyrene.

It is necessary to note that although the three types of methylcellulose were all good aminopyrene binders, they also increased the mutagenicity of aminopyrene-nitrite model (Table IV). It might be postulated that 1) the reaction rate of aminopyrene and nitrite might be too fast for the fibers to trap aminopyrene in order to retard the formation of mutagenic molecules or 2) nitrite ions could react with the scavenged molecule of aminopyrene and the molecules of nitrosated product had lesser affinity with the fiber thus they were continuously released to

Table IV. Aminopyrene binding strength of soluble fibers. Data expressed as means and standard deviations of revertant bacteria per plate (n = 4).

Fiber	Portion	TA98		TA100	
		μg fiber per plate	revertants per plate	μg fiber per plate	revertants per plate
agar	A	185	1600 (83)	462	720 (32)
	B	185	1665 (35)	462	886 (25)
	C	185	1280 (64)	462	539 (65)
guar gum	A	185	1622 (118)	462	698 (77)
	B	185	1840 (47)	462	699 (64)
	C	185	1547 (33)	462	709 (18)
gum arabic	A	185	1399 (54)	462	642 (34)
	B	185	1568 (40)	462	592 (14)
	C	185	1486 (73)	462	740 (26)
locust bean gum	A	185	1158 (115)	462	906 (62)
	B	185	1168 (54)	462	860 (28)
	C	185	1048 (30)	462	915 (25)
pectin	A	185	2335 (107)	462	1239 (73)
	B	185	2385 (35)	462	1195 (21)
	C	185	2340 (85)	462	1345 (106)
CMC	A	185	547 (49)	462	500 (60)
	B	185	636 (41)	462	467 (44)
	C	185	578 (39)	462	441 (54)
MC 25 cps	A	185	488 (36)	462	438 (29)
	B	185	264 (18)	462	266 (17)
	C	185	517 (45)	462	410 (44)
MC 1500 cps	A	185	524 (46)	462	416 (18)
	B	185	376 (27)	462	262 (21)
	C	185	644 (23)	462	413 (14)
MC 4000 cps	A	185	544 (20)	462	501 (76)
	B	185	240 (49)	462	230 (30)
	C	185	641 (55)	462	538 (16)

CMC = carboxymethylcellulose, MC = methylcellulose, cps = centipoise

*Portion A: Mixture of fiber and 0.004M aminopyrene was incubated at 37 °C for 1 h and 1 ml of the mixture was transferred to react with 0.004M sodium nitrite for 2 h.

*Portion B: Remaining mixture of fiber and 0.004M aminopyrene (after taking Portion A) was added with 1 ml of sterile-pH 3.0 water and incubated at 37 °C for another 1 h. One ml of the mixture (Portion B) was taken to react with 0.004M sodium nitrite for 2 h.

*Portion C: The second reaction mixture tube containing fiber and 0.004M aminopyrene was incubated at 37 °C for 2 h; then, 1 ml of the mixture was transferred to react with 0.004M sodium nitrite for 2 h.

interact with the tester strains of Ames test. Further experiments to obtain the kinetic parameters of the interaction between nitrite and most precursors of mutagens occurring in the gastric conditions are necessary in order to give a clearer view on how to prevent the formation of such mutagens. Therefore, from the present experiment it is possible to draw a conclusion, that in order to protect consumers from the mutagens formed from nitrite salt during the gastric digestion, the supplementary fiber must be a good, rapid nitrite scavenger and the scavenging activity should have a strong association constant.

Conclusion

The consumption of fruits and vegetables provide some nitrite scavengers and would protect the consumer from mutagens derived from nitrite treated compounds occurring during gastric digestion. However, the results obtained were contrary, when food grade soluble fibers were studied. More data on specific components of fruit and vegetable fiber binding nitrites or mutagens are required.

Literature Cited

1. Potter, J. D.; McMicheal, A. J. *J. Natl. Cancer Inst.* **1986**, *76*, 557.
2. Kinlen, L. J.; Hermon, C.; Smith P. G. *Br. J. Cancer* **1983**, *48*, 355.
3. Moller, M.E.; Dahl, R.; Bockman, O.C. (1988) *Food Chem. Toxicol.* **1988**, *26*, 841.
4. Harris, P. J.; Roberton, A. M.; Watson, M. E.; Triggs, C. M.; Ferguson, L. R. *Nutr. Cancer* **1993**, *19*, 43.
5. Eastwood, M. A.; Mitchell, W. D. In *Fiber in Human Nutrition*; Spiller ,G. A.; Amen, R. G. Eds.; Plenum Press: New York, NY, 1976; pp 118-119.
6. Takeda, Y.; Kanaya, H. *Chem. Pharm. Bull.* **1982**, *30*, 3399.
7. Kangsadalampai, K.; Juharitdamrong, S. *Thai J. Pharm. Sci.* **1996**, *20*, 107.
8. Yahagi, T.; Nagao, M.; Seino, Y.; Matsushima, T.; Sugimura, T.; Okada, M. *Mutat. Res.* **1977**, *48*:121.
9. Dence, C. W. In *Lignins Occurrence, Formation, Structure and Reactions*; Sakanen, K. V., Ludwig, C. H., Eds., Wiley-Interscience: New York, NY, 1971; pp 373.
10. Moorman, W. F. B.; Moon, N. J.; Worthington, R. E. *J. Food Sci.* **1983**, *48*, 1010.
11. Kikugawa, K.; Kato, T. In *Mutagens in food: detection and prevention;* Hayatsu, H., Ed.; CRC Press: Boca Raton, Florida, 1991, pp. 205-217.
12. Williams, G. M.; Weisburger, J. H. In *Toxicology: The basic sciences of poison;* Amdur, M. O., Doull, J., Klaassen, C. D. Eds; 4th edition; Pergamon Press: New York, N.Y., 1991, p 170.

Tea and Related Compounds

Chapter 19

Antitumor Effect of Green Tea on Rat Urinary Bladder Tumor Induced by N-Butyl-N-(4-hydroxybutyl) nitrosamine

M. Matsushima and D. Sato

Second Department of Urology, School of Medicine, Toho University, Tokyo 153, Japan

The effect of various types of tea, including leaf and powdered green tea, on N-butyl-N-(4-hydroxybutyl)nitrosamine (BBN)-induced urinary bladder carcinogenesis in rats were investigated. Green tea had an antitumorgenic effect on urinary bladder tumor induced in rats by BBN, while half-fermented tea (oolong tea) and fermented tea (black tea) did not. Powdered green tea had a concentration-dependent effect on urinary bladder tumor, and the antitumorgenic effect of powdered green tea was pronounced than that of leaf green tea.

The incidence of urinary bladder cancer in Japan is very low compared with that in the USA. However, the urinary bladder cancer rate in 3rd generation, or later, Japanese-Americans in about 3 times that of native Japanese (1). a number of contributing environmental factors has been identified in the case of bladder cancer. Of that, the role of diet is very important in urinary bladder carcinogenesis. Green tea has a long history, and its therapeutic effects have been frequently reported. Recently, attention has been paid to the anti-carcinogenic effect of its principal ingredient, cathechin (2). In the present study, we investigated the inhibitory effect of the various types of tea, including green tea, on the carcinogenesis of urinary bladder tumors induced in rats by N-butyl-N-(4-hydroxybutyl)nitrosamine (BBN).

Materials and Methods

First Experiment. Fifty-seven weeks old male Wistar rats divided into two groups, and BBN was added to their drinking water for 5 weeks at a concentration of 0.05%. The drinking water was then replaced with tap water for the control group and sencha (green tea) for the green tea group.

Second Experiment. One hundred and twenty-seven weeks old male Wistar rats were used. We divided them into 6 groups and BBN was added to their drinking water for 5 weeks at a concentration of 0.05%. At the beginning of week six, the drinking water was replaced by ordinary tap water for the control group (group 1), sencha (green tea) for group 2, macha for group 3, hojicha (roasted green tea) for group 4, oolong tea for group 5, and black tea for group 6.

Methods of Tea Preparation. Sencha and Hojicha: We poured 1200 mL of hot tap water on 10 g tea leaves of commercially available sencha and hojicha. After 60 seconds, the decoction was cooled to room temperature and given as drinking water to the animals. Macha: We poured 400 mL of hot water on 10 g of the commercially available macha. After whipping, the decoction was diluted 100 times and given as drinking water to the animals. Oolong tea and black tea: We poured 100 mL of hot tap water on a tea pack of commercially available oolong tea or black tea. After 3 min, the decoction was cooled to room temperature and given as drinking water to the animals.

Third Experiment. One hundred seven weeks old male Wistar rats were used. We divided them into five groups, and BBN was added to their drinking water at a concentration of 0.05% for 5 weeks. After that, the drinking water was replaced with tap water. The control group was fed a CE-2 pellet diet (by Nippon Clea Co.), group 2 was fed 0.15% powdered green tea CE-2 pellet diet, group 3 a 1.5% powdered green tea CE-2 pellet diet, group 4 a 3% powdered green tea CE-2 pellet diet and group 5 a 6% powdered green tea CE-2 pellet diet for 40 weeks.

Results

First Experiment. All rats were sacrificed at 40 weeks and their urinary bladders were distended with 10% formaldehyde solution and a ligature placed around the neck of each bladder to maintain proper distention. After fixation, the number of gross tumors was counted and the size of each tumor measured by ruler. The bladder was then sliced and prepared for light microscopy. Tissues were embedded in paraffin and stained with hematoxyline and eosin. Gross tumors were nearly all spherical papillary tumors like that shown in Fig. 1-3. Regarding the mean volume per tumor and total tumor volume, there were significant differences between groups. However, there were no significant differences between groups in the number of rats with tumors or the number of tumors per rat. The body weight of the rats was nearly identical in both groups, with no side effects noted. The results indicate that oral administration of green tea suppresses the growth of BBN-induced bladder tumors in rats (Fig. 4).

Second Experiment. All rats were sacrificed at 40 weeks. Regarding the number of tumors per rat, there were significant differences between the green tea and roasted tea groups and the control group. However, there were no significant differences between the macha, oolong tea, and black tea groups, and the control group. There were also significant differences in mean tumor volume per rat between the green tea and control group. The total tumor volume of the green tea group was one tenth that of the control groups. In this experiment, green tea had an antitumor effect on BBN-induced urinary bladder tumors in rats. However, the half-fermented tea (oolong tea) and fermented tea (black tea) did not have such an effect (Fig. 5).

Third Experiment. All rats were sacrificed at 40 weeks. Tumor incidence in groups 1, 2, 3, 4, and 5 was 96%95%, 75%, 88%, and 92%, respectively.

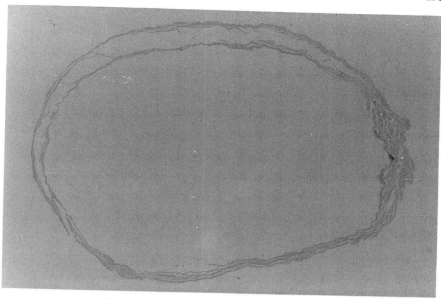

Fig. 1. Normal urinary bladder of Wistar rat.

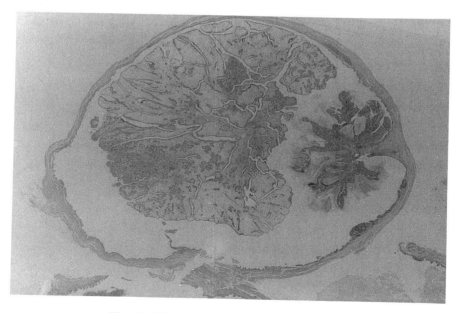

Fig. 2. Urinary bladder tumor of the control group.

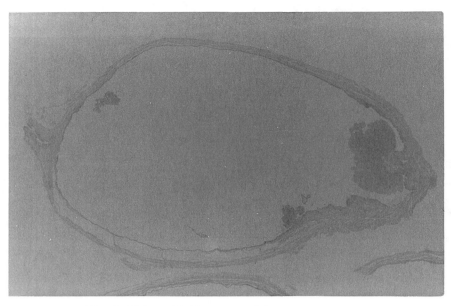

Fig. 3. Urinary bladder tumor of the green tea group.

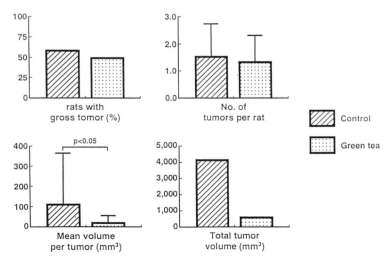

Fig. 4. Effect of green tea on BBN-induced urinary bladder carcinogenesis in rats.

Regarding the number of tumors per rat, there was significant difference between group 4 and the control group. Regarding the mean tumor volume and total tumor volume of each group, there were significant dose-dependent differences between groups 2 to 5, and the control group. In conclusion, powdered green tea had an antitumor effect on BBN-induced urinary bladder tumor in rats. The antitumor effect of the powdered green tea was stronger than that of leaves green tea (Fig. 6).

Discussion

Numbers of contributing environmental factors have been identified in the development of urinary bladder cancer. Among them, diet is very important in urinary bladder carcinogenesis (1). Investigation of dietary factors indicates a higher risk of bladder cancer due to levels of vitamin A and infrequent consumption of carrots, milk, and cruciferous vegetables, as well as too much animal meat and fat and excessive coffee consumption. Vital statistics for Japanese show that the death rate for all cancer in both males and females in Shizuoka prefecture is much lower than the Japanese average (2). The standardized mortality ratios for all cancers and for stomach cancer in the mid-western areas of Shizuoka prefecture-where green tea is a staple product- were much lower compared with the national average in both sexes (2). A significant difference in the number of habitual green tea drinkers in the town of Kawane (with a low standardized stomach cancer mortality ratio) and the towns of Osuka (with a high standardized stomach cancer mortality ratio) were observed in both sexes. The extract of fresh green tea leaves was effective in inhibiting the growth mouse sarcoma 180 (2). The result of our first experiment indicates that oral administration of green tea suppresses the growth of urinary bladder BBN-induced tumors in rats. Liberal intake of green tea and fruit juices may be responsible for low level of urinary bladder cancer in metropolitan Nagoya in Japan (3). These data suggest the possibility that green tea may reduce bladder cancer risk. Green tea leaves contain about 10% catechins, which have the following composition: (-)-epigallocatechin gallate (EGCg) 50%, (-)-epigallocatechin (EGC) 30%, (-)-epicatechin gallate (ECG 10%, and (-)-epicatechin (EC) 10%. Tea catechins are responsible for the astringent taste of green tea. Hara et al. reported the suppression of the growth of implanted sacroma 180 tumors in mice by EGCg (4). Kada et al found that EGCg has an antimutagenic effect on spontaneous reverse mutations in bacterial systems (5), while Okada et al. found that EGCg inhibited the mutagenicity of Trp-P-1 from a tryptophan pyrolysate (6). Carcinogenic mechanisms are composed of at least two steps, initiation and promotion (7). Nakamura et al found that the extracts of Japanese green tea and EGCg inhibited N-methyl-N-nitro-N-nitrosoguanidine (MNNG)-induced mutagenicity in Escherichia coli and decreased the 12-O-tetra-decanoyl-13-acetate (TPA)-mediated promotion of neoplastic transformation in mounse epidermal JB6 cells *in vitro* (8). Horiuchi et al demonstrated that EGCg also inhibited tumor promoting activity induced by teleocidine, a potent promoter, in two-stage skin carcinogenesis experiments with mice (9). These results suggest that extract of green tea leaves and EGCg may have inhibitory effects on both stages of initiation and promotion in carcinogenesis. Matsumoto et al. found that tea catechins, black tea extract and oolong tea extract have a chemopreventive addition against hepatocarcinogenesis (10). The results of our second experiment indicate that extracts of green tea had an antitumor effect on BBN-induced bladder tumor, while half-fermented tea (oolong tea) and fermented tea (black tea) did not. EGC and EGCg-the main polyphenolic constituents of green tea responsible the inhibition of tumor initiation and promotion-are present in much lower concentration in black tea and oolong tea.

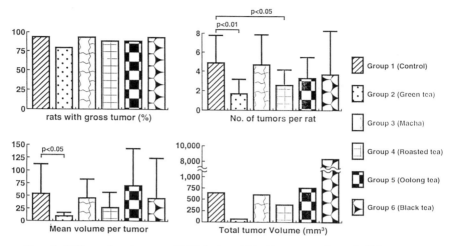

Fig. 5. Effect of each tea on BBN-induced bladder tumors in male Wistar rats.

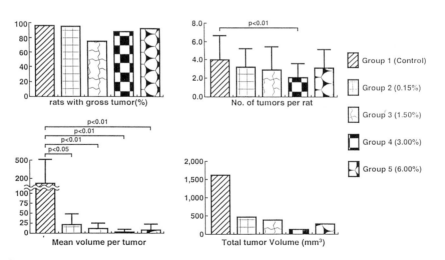

Fig. 6. Effect of powdered green tea on BBN-induced bladder tumors in male Wistar rats.

Most Japanese drink green tea brewed from green tea leaves, discarding the tea grounds. Many nutrients, including vitamin A, vitamin E, chlorophyll, food fiber and lipid, are still contained in the tea groups. Recently powdered green tea has become available in Japan, allowing drinkers to consume all the dried tea matter in beverage form. In our third experiment, we investigated the antitumor effect stronger than that of the green tea extracts.

The amount of green tea, roasted tea, oolong tea and black tea administrated in this study corresponds to 6 to 7 cups of each tea in humans, and 1.5% dose of powdered green tea administered in this study corresponds to about 10 cups of green tea in humans. Since the present results suggest that green tea extract and powdered green tea eating may protect against human urinary bladder carcinogenesis.

Acknowledgments

This work was supported in part by a Grant-in Aid from the ministry of Education, Science, Sports and Culture, Japan. The authors would like to thank Kiyoshi Tajima for animal care.

Literature Cited

1. Hueper, W. G. *Occupational and Environmental Cancers of the Urinary System.* Yale University Press, New Haven, 1969.
2. Oguchi, I.; Nasu, K.; Kanaya, S.; Ota, Y.; Yamamoto, S.; Nomura, T. Epidemiological and experimental studies on the antitumor activity by green tea extracts. *Jpn. J. Nutr.* **1989,** *47,* 93-102.
3. Ohno, Y.; Aoki, K.; Obata, K.; Morrison, S. Case-control study of urinary bladder cancer in metropolitan Nagoya. *Nat. Can. Inst. Monograph* **1985,** *69,* 229-234.
4. Hara, M.; Nakamura, K.; Fujino, R.; Hosaka, H.; Kohisae, S.; Asai, H.; Sugiura, M. Abstracts of Papers, *The 43rd Annual Meeting of Japanese Cancer Assoc.:* 1984, p 993.
5. Kada, T.; Inoue, T.; Sadaie, Y.; Shimoi, K. Abstracts of Papers, *The Annual Meeting of the Agricultural Chemical Society of Japan,* Kyoto, April, 1981, p 10. and Kada, T.; Kaneko, K.; Matsuzaki, S.; Hara, Y. *Mutation Res.* **1985,** *150,* 127-132.
6. Okuda, T.; Mori, K.; Hayatsu, H. *Chem. Pharm. Bull* **1984,** *32,* 3755-3758.
7. Berenblum, L. *Cancer Res.* **1941,** *1,* 44-48.
8. Nakamura, Y.; Harada, S.; Hara, S.; Tomita, I. Abstracts of Papers, *The 15th Annual Meeting of the Environmental Mutagnic Society of Japan,* Tokyo, 1986, p 50.
9. Horiuchi, T.; Fujiki, H.; Yoshizawa, S.; Okuda, T.; Sugimura, T. Abstracts of Papers, *The 45th Anjual Meeting of the Japanese Cancer Assoc.,* Sopporo, 1986, p 81.
10. Matsumoto, N.; Kohri, T.; Okushio, K.; Hara, T. Inhibitory effects of tea catechin, black tea extract and oolong tea extract on hepatocarcinogenesis in rat. *Japn. J. Cancer Res.* **1996,** *87,* 1034-1038.

Chapter 20

In Vitro and In Vivo Effects of Fatty Acids and Phenolic Compounds on Chemical Mediator Release from Rat Peritoneal Exudate Cells

K. Yamada[1], N. Matsuo[1,3], K. Shoji[1], M. Mori[1], T. Ueyama[1], S. Yunoki[1], S. Kaku[1], S. Oka[2], K. Nishiyama[2], M. Nonaka[1], H. Tachibana[1], and M. Sugano[1,4]

[1]Department of Food Science and Technology, Kyushu University, Fukuoka 812-81, Japan
[2]Department of Biological Resource Sciences, Miyazaki University, Miyazaki 889-21, Japan

Phenolic antioxidants in foodstuffs, such as tea polyphenols and flavonoids, inhibited histamine and LTB_4 release from rat peritoneal exudate cells. Among them, triphenol compounds suppressed histamine release, but not diphenol compounds. In LTB_4 release, most phenolic compounds with 1,1-diphenyl-2-picrylhydrazyl radical scavenging activity exerted an inhibitory effect. Carboxylated phenols, or quercetin glycosides, exerted weak activity, irrespective of their strong radical scavenging activity. This suggests that membrane permeation of phenols is necessary for the inhibition of LTB_4 release. In addition, unsaturated fatty acids with more than 3 double bonds significantly suppressed LTB_4 release. Feeding experiments suggested that perilla oil, rich in n-3 α-linolenic acid, decreases LTB_4-releasing activity of the cells, probably by reducing the proportion of arachidonic acid in membrane phospholipids. Tea polyphenols feeding decreased the activity through some different mechanism.

Allergic reactions are usually classified into 4 types. Type I allergy plays an important role in the induction of allergies against food proteins, or airborne antigens (*1*). In the expression of the type I allergy, induction of allergen-specific IgE, and release of chemical mediators such as histamine and leukotrienes (LT) from mast cells or basophils, are critical steps. Recently, it has been reported that some food components affect the above immune processes. For example, unsaturated fatty acids (UFA) seem to enhance type I allergic reactions through enhancement of IgE production, and inhibition of IgA and IgG production by rat mesenteric lymph node lymphocytes (*2*). Since the expression of IgE production-enhancing activity of UFA was inhibited in the presence of a lipophilic antioxidant α-tocopherol, lipid peroxidation seems to enhance type I allergy. On the other hand, natural antioxidants such as flavonoids (*3, 4*) and tea polyphenols (TP) (*5, 6*) have been

[3]Current address: Department of Biochemistry, Oita Medical University, Oita 879-55, Japan.
[4]Current address: Faculty of Human Health Science, Prefectural University of Kumamoto, Kumamoto 862, Japan

reported to suppress type I allergy through the inhibition of chemical mediator release. These findings suggest that various types of food components can regulate allergic reactions. Thus, clarification of a mechanism by which they regulate allergic reactions may have great benefits in the prevention of allergies. However, information on the regulatory mechanism is so far, limited. Thus, we studied the structure-activity relationship on the chemical mediator release inhibitory effect of phenolic compounds.

In addition, polyunsaturated fatty acids (PUFA) have been reported to regulate various biological functions including immune reactions (7-11). Among them, arachidonic acid (n-6) is a substrate of lipoxygenase, and gives 4-series LT which induces the type I allergy. On the contrary, another n-6 γ-linolenic acid (12, 13), and n-3 PUFA, such as eicosapentaenoic (EPA) and docosahexaenoic acids (DHA), have been reported to be anti-allergic (7, 9). We also studied on the effect of PUFA on chemical mediator release from rat peritoneal exudate cells (PEC) to clarify the mechanism of their anti-allergic effect and found that PUFA, with more than 3 double bonds, strongly suppressed LTB_4 release from the cells (11). However, the mechanism by which PUFA suppress LTB_4 release was not clarified well, especially *in vivo*. Thus, we studied the effect of dietary fats, and TP, on the chemical mediator releasing activity of PEC.

In Vitro Assay System for Chemical Mediator Release-inhibitory Substances

Preparation of Rat Peritoneal Exudate Cells. PEC were isolated by the method of Matsuo et al. (5) from 8-9 weeks old male Wistar rats (Ceac Yoshitomi, Fukuoka, Japan). At first, Tyrode buffer (137 mM NaCl, 2.7 mM KCl, 1.8 mM $CaCl_2$, 1.1 mM $MgCl_2$, 11.9 mM $NaHCO_3$, 0.4 mM NaH_2PO_4, 5.6 mM glucose, pH 7.2), supplemented with 0.1% bovine serum albumin was injected into the peritoneal cavity. The abdomen was then massaged gently for 2 min. After opening the peritoneal cavity, the fluid containing PEC was recovered, and the cells were washed once with Tyrode buffer. After centrifugation at 200 x g for 10 min at 4°C, cells were resuspended in a modified-ammonium chloride buffer (150 mM NH_4Cl, 10 mM $KHCO_3$, 10 mM EDTA-2Na, pH 7.4) and then incubated for 5 min at 4°C. After centrifugation, cells were resuspended in Tyrode buffer, supplemented with 0.9 mM $CaCl_2$.

Determination of histamine. After adjusting the cell density at 0.5 x 10^6 cells/ml, cells were stimulated with 5 μM A23187 for 20 min at 37°C, in the absence or presence of phenolic compounds. DMSO concentration of the cell suspension was adjusted to the levels below 1%, where the solvent exerted no effect on the mediator release. The stimulation was terminated by incubating the cells for 15 min at 4°C and the cell suspension was centrifuged for 10 min at 300 x g.

Histamine content in the supernatant was measured according to the method of Shore *et al.* (14), with slight modifications (5). After mixing 2 ml of the above supernatant with 0.75 g of NaCl and 0.5 ml of 1 N NaOH, 5 ml of the 3:2 (v/v) mixture of *n*-butanol and chloroform was added. After mixing for 5 min, the solution was centrifuged for 5 min at 270 x g and the upper butanol layer was recovered. Four ml of the solution was mixed with 2 ml of n-heptane and 1.5 ml of 0.1 N HCl for 5 min. After centrifugation for 5 min at 270 x g, 1 ml of the HCl layer was recovered, and mixed with 0.15 ml of 1 N NaOH and 0.1 ml of 0.2% *o*-phthalaldehyde. This solution then stood for 5 min at room temperature. The reaction was terminated by adding 0.14 ml of 0.5 N H_2SO_4, and fluorescence intensity was measured using a spectrophotometer (type RF500, Shimadzu, Kyoto, Japan), with the excitation at 360 nm, and the emission at 450 nm. Since around 20% of intracellular histamine was released spontaneously without A23187 stimulation, relative histamine release (%) was calculated according to the following formula: (A23187-stimulated release with test sample - spontaneous release/A23187-stimulated release without test sample - spontaneous release) x 100.

Determination of LTB4. LT released into Tyrode buffer after an A23187 stimulation was measured, according to the method of Powell (15), with slight modifications (5). PEC (2 x 10^6 cells) were suspended in 40 µl of Tyrode buffer containing 0.9 mM $CaCl_2$ at 4, mixed with 5 µl of sample solution and 5 µl of 50 µM A23187, and then incubated for 20 min at 37. The reaction was terminated by adding 50 µl of the 30:25 (v/v) mixture of acetonitrile and methanol. After adding 50 ng of PGB_2 as the internal standard, the cell suspension was centrifuged for 10 min at 300 x g. The supernatant was filtered through a 0.22 µm nitrocellulose filter (Millipore, Tokyo, Japan). The filtrate (20 µl) was applied to a ODS-A column (150 x 6 mm, 5 µm particle size) attached to the Shimadzu SCL-10A high performance liquid chromatography system. Eicosanoids were eluted from the column with a 30:25:45 (v/v/v) mixture of acetonitrile, methanol and H_2O containing 5 mM CH_3COONH_4 and 1 mM EDTA-2Na (pH 5.6), at a flow rate of 1.1 ml/min. PGB_2 and LTB4 were detected by the absorbance at 280 nm. Relative LTB4 release (%) was calculated according to the following formula: (A23187-stimulated release with test sample/A23187-stimulated release without test sample) x 100.

Determination of DPPH radical-scavenging activity. Polyphenols, or antioxidants, were dissolved in ethanol to measure their 1,1-diphenyl-2-picrylhydrazyl (DPPH) radical-scavenging activity. Reaction mixtures containing 20 mM HEPES (pH 7.4), 0.05 mM DPPH, 60% ethanol and 0.05 mM test samples were stood for 30 min at room temperature in dark place and then absorbance at 517 nm was measured. DPPH radical scavenging activity (%) was calculated according to the following formula: (A$_{517}$ without test sample - A$_{517}$ with test sample/A$_{517}$ without test sample) x 100.

Statistics. When necessary, data was analyzed by Duncan's new multiple range test to evaluate significant differences (16).

***In Vitro* Effect of Food Components on Chemical Mediator Release from Rat Peritoneal Exudate Cells**

Effect of unsaturated fatty acids. When rat PEC were stimulated with A23187 in the presence of 0.1 mM UFA, oleic (18:1n-9) and linoleic (18:2n-6) acids did not affect LTB4 release from the cells. Whereas, fatty acids with more than 3 double bonds exerted a significant inhibitory effect (11). The inhibitory effect was enhanced with the number of double bonds in the order of α-linolenic acid (18:3n-3) < γ-linolenic acid (18:3n-6) < EPA (20:5n-3) < DHA (22:6n-3). The inhibitory effect was detected only at 100 µM in arachidonic acid, above 10 µM in EPA, and above 1 µM in DHA. Since production of LTB5 was observed only in the presence of EPA, at least a part of extracellular PUFA may be incorporated into the cells and oxidized with lipoxygenase. LTB5 is produced by lipoxygenase oxidation of EPA, and reported to suppress the production of LTB4. However, arachidonic acid, the precursor of LTB4 suppressed LTB4 release. In addition, DHA, which is not metabolized to LT, exerted the highest suppressing effect. These results suggest that PUFA directly suppress LTB4 production, probably through inhibition of the lipoxygenase activity.

On the other hand, these PUFA have also been reported to enhance type I allergy through stimulation of IgE production, and inhibition of IgA and IgG production by rat mesenteric lymph node lymphocytes (2). Since the expression of their IgE production enhancing activity was inhibited in the presence of lipophilic antioxidants, such as α-tocopherol, oxidation of PUFA is responsible for the process. Fortunately, various antioxidants have been reported to act as an anti-allergic agent by inhibiting chemical mediator release (5, 6). These results suggest that combinational use of lipophilic antioxidants is essential to express anti-allergic effect of PUFA, without expression of IgE production-enhancing activity

Effect of tea polyphenols. TP have been reported to exert diverse biological effects, such as anti-oxidative, anti-bacterial, anti-fungal, anti-viral, anti-atherosclerotic and anti-cancer activities (17-20). In addition, TP with the triphenol group, such as epigallocatechin (EGC), epicatechin gallate (ECg) and epigallocatechin gallate (EGCg) have been reported to suppress both histamine and LTB4 release from rat PEC, while TP with the diphenol group, such as catechin and epicatechin, do not (5). Similarly, triphenolic TP have been reported to exert transformed cell-specific toxicity, but not diphenolic TP (20). In the case of immunoglobulin production, triphenolic TP suppressed IgE production of rat mesentric lymph node lymphocytes bellow 0.1 mM, but diphenolic TP did not (21). These results suggest that the triphenol group has a stronger biological effect than the diphenol group.

TP inhibits histamine release from rat basophilic RBL-2H3 cells stimulated with a nonspecific stimulant A23187, and with antigen-IgE complex composed of anti-DNP IgE and DNP-BSA (6). In addition, they inhibited both histamine and LTB4 release from mast cells present in rat PEC. Histamine is accumulated in secretory granules, and released from the cells after an elevation of intracellular Ca^{2+} level, and the activation of protein kinase C (22, 23). On the other hand, LT are released after the elevation of the Ca^{2+} level, activation of phospholipase A2 (PLA2), and lipoxygenase oxidation of arachidonic acid released from membrane phospholipids by PLA2 (24,25). Simultaneous suppression of histamine and LTB4 release suggests that TP exert their effect through a common reaction, such as the elevation of the intracellular Ca^{2+} level. However, the increase of Ca^{2+} occurred irrespective of the presence of EGCg (6). Thus, TP may inhibit the process after the elevation of Ca^{2+}.

Effect of simple polyphenols. In addition to TP, simple polyphenols can suppress histamine release from RBL-2H3 cells (5). In this case, the inhibitory effect of diphenolic compounds such as catechol, resorcinol and hydroquinone was negligible. Whereas, triphenolic compounds such as pyrogallol, and gallic acid, strongly suppressed histamine release, as seen in TP. A similar result was obtained in rat PEC, stimulated with A23187, in the presence of simple polyphenols. As shown in Table I, triphenols, such as pyrogallol and gallic acid, inhibited the release of histamine; mono- and di-phenolic compounds did not. These results suggest that the triphenol structure is essential for the expression of their histamine release inhibitory activity.

Table I. Chemical Mediator Release Inhibitory and DPPH Radical Scavenging Activities of Phenolic Compounds

Phenolic Compounds	Relative histamine release (%)	Relative LTB4 release (%)	DPPH radical scavenging activity (%)
None	100 ± 4^a	100 ± 4^a	0
Catechol	102 ± 1^a	27 ± 6^b	82.9
Resorcinol	134 ± 2^b	92 ± 12^a	19.8
Hydroquinone	121 ± 2^c	45 ± 4^b	84.8
Pyrogallol	73 ± 1^d	22 ± 2^b	82.2
Salicylic acid	116 ± 8^c	94 ± 13^a	0
Protocatechuic acid	125 ± 2^{bc}	76 ± 5^a	83.1
Gallic acid	87 ± 3^c	99 ± 12^a	82.3

Data are mean±SE (n=3) and values not sharing a common letter are significantly different at $p < 0.05$.

On the other hand, a different structure-activity relationship was observed in LTB4 release. As shown in Table I, *o*-diphenolic catechol, and *p*-diphenolic hydroquinone, exerted a strong inhibitory effect as well as pyrogallol, but *m*-diphenolic resorcinol did not. The LTB4 release inhibiting activity of simple polyphenols is related to their DPPH radical-scavenging activity. Thus, the inhibition of LTB4 release seemed to be related to their anti-oxidative activity. In the case of carboxylated phenols, any compounds such as monophenolic salicylic, diphenolic protocatechuic, and triphenolic gallic acids, could not inhibit LTB4 release irrespective of their strong DPPH radical scavenging activity. Kohiyama *et al.* (*26*) also reported that carboxylated diphenols strongly inhibited the lipoxygenase reaction in a cell free system, but not in intact rat platelets. In addition, a carboxylated catecholamine has been reported to lose cellular DNA breaking activity, though it exerts DNA breaking activity *in vitro,* as strong as those without a carboxyl group (*27*). The addition of a highly hydrophilic carboxyl group may reduce the membrane permeability of phenolic compounds. The inability of carboxylated compounds suggests that penetration of polyphenols through cell membranes is essential for the inhibition of LTB4 release. Since a carboxylated triphenol, gallic acid, inhibited histamine release as strongly as pyrogallol, triphenols may interact with cell surface component(s) to suppress histamine release.

Effect of flavonoids. Flavonoids are another group of natural phenolic antioxidants which have been reported to inhibit chemical mediator release (*28-33*). To clarify the mechanism by which they regulate chemical mediator release, we studied here the structure-activity relationship of flavonoids on the inhibition of chemical mediator release. At first, the activity of flavonols with different B rings was examined. As shown in Table II, histamine release was weakly inhibited by the diphenolic quercetin, and strongly by triphenolic myricetin, but not by monophenolic kaempferol. This suggests that the triphenol structure in the B ring is essential for the inhibition of histamine release. On the other hand, LTB4 release was inhibited by all flavonols employed here, irrespective of the number of OH groups in the B ring. Thus, some structure other than the B ring, appeared to be participated in the inhibition of LTB4 release.

Table II. Chemical Mediator Release Inhibitory Activity of Flavonoids and Quercetin Glycosides

Flavonoids	Sugar	Relative histamine release (%)	Relative LTB4 release (%)	DPPH radical scavenging activity (%)
None	-	100±4[a]	100±4[a]	0
Kaempferol	-	122±6[b]	3±2[b]	83.4
Quercetin	-	81±3[c]	0[b]	83.4
Quercitrin	Rhamnose	Not tested	55±6[c]	Not tested
Rutin	Rutinose	Not tested	73±4[d]	80.7
Myricetin	-	52±1[d]	12±0[b]	77.8
Luteolin	-	Not tested	0[b]	80.0
Cyanidin	-	Not tested	67±7[d]	48.7

Data are mean±SE (n=3) and values not sharing a common letter are significantly different at $p < 0.05$.

Since the A ring of flavonoids is m-diphenol, which exerted no regulatory activity in the case of simple polyphenols, the contribution of the C ring was further studied. When PEC were stimulated, in the presence of quercetin glycosides with various sugars at the C-3 position, the inhibitory effect of glycosides was much weaker than that of their aglycon quercetin. Since rutin exerted DPPH radical scavenging activity as strong as quercetin, the decrease of inhibitory activity may be due to the decrease of membrane permeability induced by the addition of hydrophilic sugar molecules.

Then, the activity of a flavonol (quercetin), flavone (luteolin) and anthocyanidin (cyanidin), with a diphenolic B ring was compared, to examine the importance of C3-OH and C4-carbonyl groups. Quercetin has both groups, luteolin has only a C4-carbonyl group, and cyanidin has only a C3-OH. As shown in Table II, quercetin and luteolin exerted strong inhibitory effects but the effect of cyanidin was much weaker than those of the former. DPPH radical-scavenging activity of flavonoids exerted a similar tendency with their LTB4 release inhibitory effect. These results suggest that the C4-carbonyl is essential for the expression of a strong LTB4 release inhibitory effect of flavons and flavonols. Though cyanidin exerted weaker DPPH radical scavenging activity at 50 μM, it scavenged the radical as effective as other flavonoids at 10 μM (scavenging rate 76.8%). Since the cyanidin solution is obviously colored at 50 μM, the DPPH determination seemed to be complicated by the absorption by cyanidin itself. These observations suggest that anthocyanidins have DPPH radical scavenging activity as strong as other flavonoids.

In flavonoids, three structural groups are important determinants for radical scavenging, and/or antioxidative potential: 1) the o-dihydroxy (catechol) structure in the B ring, 2) the 2,3 double bond adjacent to C4-carbonyl group, and 3) 3- or 5-OH group which interact with C4-carbonyl group (34). According to the above results, strong DPPH radical scavenging activity can be expressed only by a catechol structure in the B ring. However, the structures conjugated with a 4-carbonyl group seems to be essential for the expression of a strong LTB4 releasing inhibitory activity.

Effect of antioxidants and scavengers of reactive oxygen species in the presence of polyphenols. Phenolic compounds exert reducing activity against oxidized compounds, and scavenging activity against reactive oxygen species (ROS). To show which activity participates in the inhibition of LTB4 release, the effect of some antioxidants and ROS scavengers on LTB4 release was examined in the presence, or absence of phenolic compounds.

Table III. DPPH Radical Scavenging and LTB4 Release Inhibiting Activity of Antioxidants and Reactive Oxygen Scavengers

Additives	DPPH radical scavenging activity (%)	Relative LTB4 release (%)	
		None	Catechol
None	0	100 ± 2^a	41 ± 1^a
Butyrated hydroxytoluene	82.7	41 ± 2^b	22 ± 3^b
α-Tocopherol	82.0	71 ± 1^c	11 ± 4^c
Mannitol	0	108 ± 3^{ad}	43 ± 3^a
Triethylenediamine	0	119 ± 8^d	37 ± 3^a

Data are mean±SE (n=3) and values not sharing a common letter are significantly different at $p < 0.05$.

Butylated hydroxytoluene is a synthetic antioxidant, and α-tocopherol is a representative natural antioxidant known as vitamin E. Both antioxidants exerted DPPH radical scavenging activity as strong as catechol. On the other hand, a hydroxyl radical scavenger mannitol, and a singlet oxygen scavenger triethylene-diamine, did not exert DPPH radical scavenging activity. In addition, LTB4 release inhibitory activity was observed only in the antioxidants which exerted DPPH radical scavenging activity. The inhibitory effect of catechol was especially enhanced in the presence of α-tocopherol. On the other hand, mannitol and triethylenediamine did not affect the inhibitory effect of catechol. These results suggest that the LTB4 releasing inhibitory effect of phenolic compounds is related to their DPPH radical scavenging activity. Since radical scavengers which erase hydroxyl radical and singlet oxygen, did not suppress the expression of LTB4 release inhibitory activity of catechol, these ROS may not be participated in the inhibitory reaction.

***In Vivo* Assay System for Chemical Mediator Release Inhibitory Substance**

The above cellular experiments suggest that PUFA, and phenolic antioxidants, have anti-allergic effects through the inhibition of chemical mediator release. However, the activity observed in cellular experiments may not be always reproducible on an animal level. Thus, a feeding experiment is essential for the assessment of an anti-allergic effect of food components. In the case of PUFA, dietary EPA and/or DHA have been reported to affect eicosanoid production (*7, 9, 10*). In the case of phenolic compounds, dietary TP have been reported to affect the development of various diseases (*19*). Thus, the effect of dietary fats and TP on the chemical mediator releasing activity of rat PEC was examined.

To examine the effect of food components on the chemical mediator releasing activity of mast cells, PEC were isolated from Wistar rats after 3 weeks feeding of the diets containing the various types of dietary fats and TP, stimulated with 5 μM A23187. Then, the amounts of histamine and LTB4 were determined, according to the method of Matsuo *et al.* (*5*). Safflower oil (SA) rich in n-6 linoleic acid, perilla oil (PE) rich in n-3 α-linolenic acid, palm oil (PA) rich in saturated palmitic, and monounsaturated oleic acids were added to the diet at the 10% level. TP is a mixture of tea catechins prepared from green tea, and widely used for food additives in Japan (Taiyo Chemical Co., Yokkaichi, Japan). TP was added to the diet at the 1% level in the first feeding experiment. Body weight and food intake were measured every day. After isolating PEC, various tissues such as heart, liver, lung, kidney, spleen, and adipose tissues were excised out and weighed. Lipids were extracted according to the method of Folch *et al.* (*35*), and fatty acid composition of the phospholipid fraction was analyzed using gas-liquid chromatography according to the method of Ikeda *et al.* (*36*).

In the second feeding experiment, TP was added to the diet at 0, 0.01, 0.1 and 1.0% levels and SA which exerted no LTB$_4$ release inhibitory effect was added at the 10% level, to examine the dose-dependent effect of TP. After 3 weeks feeding, PEC are isolated to compare LTB$_4$ releasing activity and fatty acid composition, as described above.

Dietary Effect of Food Components on Chemical Mediator Releasing Activity of Rat Peritoneal Exudate Cells

Combinational effect of dietary fats and tea polyphenols. Rats ate diets containing TP, as much as diets without TP, but weight gain was lower in the TP-fed groups than in the TP-free groups. This suggests that the feeding of such high dose of TP disturbs the growth of rats. In this growing condition, histamine releasing activity of PEC was significantly higher in the PA-fed group, than in the SA- and PE-fed groups. However, TP feeding did not affect the histamine releasing activity. On the other hand, LTB$_4$ releasing activity was significantly lower in the PE-fed group than in the other groups. In addition, TP feeding significantly decreased LTB$_4$ releasing activity in all dietary fat groups.

In vitro cellular experiments suggested that triphenolic compounds interact with cell surface components, and need not be incorporated into cells to express the histamine release inhibitory effect. This means that fairly high amounts of TP should be present in extracellular fluid during the stimulation of mast cells. Washing of cells before A23187 stimulation erases the effect of extracellular TP. When catechin was orally administered to men at a dose of 92.3 mg/kg, increase in blood phenol level was 4 to 21 µg/ml, which is corresponding to 14-72 µM, during 48 hr after the single administration (*37*). Chemical mediator release inhibitory effect of TP was detectable at the doses above 10 µM in EGCg, and at 100 µM in other phenolic compounds (*5*). Though exact levels of these catechins are not reported in normal dietary conditions, EGCg may suppress histamine release *in vivo*.

Table IV. Effect of Dietary Fats and Tea Polyphenol on Chemical Mediator Releasing Activity of Rat Peritoneal Cells

Diet	Histamine releasing activity (ng/10^6 cells)	LTB$_4$ releasing activity (ng/2 x 10^6 cells)
SA	611±21[a]	59.9±1.9[a]
PE	573±39[a]	27.2±0.6[b]
PA	760±15[b]	55.4±2.2[a]
SA + TP	626±20[a]	22.4±1.8[bc]
PE + TP	613±37[a]	17.2±1.6[c]
PA + TP	827±23[b]	40.2±1.7[d]

Data are mean±SE(n=3). Values not sharing a common letter are significantly different at $p<0.05$. SA(safflower oil), PE(perrila oil), PA(palm oil), TP(tea polyphenol).

Table V. Effect of Dietary Fats and Tea Polyphenol on Fatty Acid Composition in Phospholipid of Rat Peritoneal Cells

Diet	Fatty acid (%)					
	16:0	18:0	18:1n-9	18:2n-6	20:4n-6	22:6n-3
SA	22.4±1.0	20.7±2.0	12.5±2.1^a	11.9±0.6^{ab}	20.9±1.3^a	3.0±0.2^{ac}
PE	22.6±0.3	22.3±1.0	16.0±0.4^{ab}	9.3±0.4^a	8.7±1.0^b	5.7±0.6^b
PA	21.7±1.0	16.5±0.8	17.4±1.3^b	13.0±0.6^b	14.1±2.4^c	4.6±0.9^{ab}
SA + TP	21.3±0.9	20.4±2.2	15.1±0.5^{ab}	13.1±1.1^b	14.2±0.8^c	1.7±0.2^c
PE + TP	21.2±1.1	20.3±0.9	16.3±0.9^{ab}	12.8±0.8^{bc}	7.2±0.7^b	4.2±0.8^{ab}
PA + TP	18.1±2.9	18.1±2.8	14.6±1.1^{ab}	9.6±1.9^{ac}	14.2±2.0^c	4.4±0.6^{ab}

Data are mean±SE (n=3) and values not sharing a common letter are significantly different at $p < 0.05$. SA; safflower oil, PE; perrila oil, PA; palm oil, TP; tea polyphenol.

Since LTB$_4$ is produced by lipoxygenase oxidation of arachidonic acid derived from cell membrane phospholipids, fatty acid composition of the phospholipid fraction of PEC was measured. As shown in Table V, fatty acid composition was significantly affected in 18:1n-9, 18:2n-6, 20:4n-6, and 22:6n-3. A most marked change was observed in the proportion of arachidonic acid (20:4n-6). The proportion of arachidonic acid in the PE-fed group, was almost one third of that the SA-fed group and the level of the PA-fed group was significantly lower than that of the SA-fed group. Such a decrease in the proportion of arachidonic acid may lead to the decrease of LTB$_4$ production. However, the difference in the proportion of arachidonic acid between the SA- and PA-fed groups did not reflect on the LTB$_4$ releasing activity. Since the percentage of mast cells in PEC preparations is only 10 to 30%, comparison of fatty acid composition of PEC may be somewhat misleading.

TP feeding significantly decreased the level of arachidonic acid in the SA-fed group to the level encountered in the PA-fed group, where the addition of TP did not lower it. This suggests that TP-feeding decreases the LTB$_4$ releasing activity by decreasing the proportion of arachidonic acid as shown in feeding PE. However, TP did not affect the proportion of arachidonic acid in PE- and PA-fed groups. These results suggest that PE feeding decreases the LTB$_4$ releasing activity of the cells through the reduction of arachidonic acid content, but the other mechanism is also working in the decrease of LTB$_4$ releasing activity by TP feeding.

Dose-dependent effect of dietary tea polyphenol. The TP dose employed in the first feeding experiment was the maximal dose which did not significantly decrease food intake of rats. However, weight gain was significantly decreased at this TP level. In addition, the TP level is corresponding to approximately 60 g for 60 kg person per day, and several hundred cups of green tea should be consumed every day to take this TP level. Thus, we examined the effect of TP at lower doses.

Weight gain during 21 days feeding was unchanged at the 0.01% TP level (132±2 g) compared with the control group (131±4 g), slightly decreased at the 0.1% TP level (127±1 g), and significantly decreased at the 1% TP level (101±4 g). However, a significant decrease in food efficiency was observed only in the 1% TP group (0.31±0.01 g weight gain/g intake in the 1% TP fed group vs. 0.38±0.01 g weight gain/g intake in the control group). Decrease in tissue weight was most marked in adipose tissue, and then spleen and heart, at the 1% TP level. On the other hand, a significant increase in tissue weight was observed in liver (11.6±0.4 g in control) at the 0.1% (17.5±0.4 g), and 1% (13.7±0.2 g) levels. In the kidney the levels were(2.36±0.04 g in control) at the 0.01% (2.59±0.06 g) and 0.1% (2.63±0.07 g) levels. In the spleen, the levels were (0.71±0.05 g in control) at the 0.1% level (0.96±0.04 g). These results suggest that TP dose around 0.01% is a safety level which does not largely affect tissue weight.

Table VI. Dose-dependent Effect of Dietary Tea Polyphenol on Chemical Mediator Releasing Activity and Proportion of Arachidonic Acid in Phospholipids of Rat Peritoneal Cells

TP dose (%)	Histamine release (%)	LTB$_4$ releasing activity (ng/2 x 10^6 cells)	Arachidonic acid (%)
0	94.6 ± 1.6^a	3.67 ± 0.50^a	16.4 ± 1.2^a
0.01	87.6 ± 0.7^b	1.63 ± 0.30^b	18.0 ± 1.0^a
0.1	81.1 ± 1.1^c	1.98 ± 0.18^b	18.8 ± 0.3^a
1.0	90.1 ± 2.0^b	1.44 ± 0.27^b	13.2 ± 1.1^b

Data are mean±SE (n=6) and values not sharing a common letter are significantly different at $p < 0.05$.

For chemical mediator release, a significant decrease in histamine releasing activity was observed in the groups fed 0.01 and 0.1% TP, though the effect was very weak (Table IV). On the contrary, a decrease in the LTB$_4$ releasing activity was more marked and observed even at 0.01% level. Though the proportion of arachidonic acid was decreased in the 1% TP fed group, it was increased in the 0.01 and 0.1% TP-fed groups. However, a decrease in the LTB$_4$ releasing activity was observed in all TP doses. This suggests that the decrease in LTB$_4$ releasing activity was not necessarily induced by the decrease in the proportion of arachidonic acid in the phospholipid fraction of PEC.

The lowest TP dose, 0.01%, is corresponding to the intake of approximately 600 mg for 60 kg person per day. Since a cup of green tea contains approximately 100 mg of TP, several cups of green tea per day is enough to take the level of TP. Thus, daily consumption of various types of teas containing TP may suppress the allergic reactions in allergic patients, as well as reduction of cancer risk (19). Such anti-allergic effects of TP may be enhanced by the other food components, such as dietary fats rich in n-3 PUFA, as shown here. In addition, it is reported that dietary α-tocopherol and sesamin did not exert a significant effect on LTC$_4$ productivity of rat lung tissue in individual feeding, but simultaneous feeding dramatically decreased LTC$_4$ productivity in a lung tissue-specific manner (38). These results suggest that synergism of food components should be clarified to construct designer foods with multifunctional activities.

Acknowledgment

This work was partly supported by Special Coordination Funds for Promoting Science and Technology of the Science and Technology Agency of Japan.

Literature Cited

1) Metcalfe, D.D. *Current Opinion Immunol.* **1991,** 3, 881-886.
2) Yamada, K.; Hung, P.; Yoshimura, K.; Taniguchi, S.; Lim, B. O.; Sugano, M. *J. Biochem.* **1996**, 120, 138-144.
3) Baumann, J; Bruchhausen, F.; Wurm, G. *Prostaglandins* **1980**, 20, 627-639.
4) Corvazier, E.; Maclouf, J. *Biochim. Biophys. Acta* **1985**, 835, 315-321.
5) Matsuo, N.; Yamada, K.; Yamashita, K.; Shoji, K.; Mori, M.; Sugano, M. (1996) *In Vitro Cell. Develop. Biol.* **1996**, 32A, 340-344.
6) Matsuo, N.; Yamada, K.; Shoji, K.; Mori, M.; Sugano, M. *Allergy* **1997**, 52, 58-64.
7) Lee, T. H.; Hoover, R. L.; Williams, J.D.; Sperling, R. I.; Ravalese III, J.; Spur, B. W.; Robinson, D. R.; Corey, E. J.; Lewis, R.A.; Austen, K. F. *New Engl. J. Med.* **1985**, 312, 1217-1224.

208

8) Prescott, S. M. *J. Biol. Chem.* **1984**, 259, 7615-762.
9) Terano, T.; Salmon, J. A.; Moncada, S. *Biochem. Pharmacol.* **1984**, 33, 3071-3076.
10) Thien, F. C. K.; Hallsworth, M. P.; Soh, C.; Lee, T. H. *J. Immunol.* **1993**, 150, 3546-3552.
11) Yamada, K.; Mori, M.; Matsuo, N.; Shoji, K.; Ueyama, T.; Sugano, M. *J. Nutr. Sci. Vitaminol.* **1996**, 42, 301-311.
12) Schalin-Karrila, M.; Mattila, L.; Jansen, C. T.; Uotila, P. *Br. J. Dermatol.* **1987**, 117, 11-19.
13) Lovell, C.R.; Burton, J. L.; Horrobin, D.F. *Lancet* **1981**, i, 278-279.
14) Shore, P.A.; Burkhalter, A.; Cohn, V. H. *J. Pharmacol. Exp. Therapy* **1959**, 127, 182-186.
15) Powell, W. S. *Anal. Biochem.* **1987**, 164, 117-131.
16) Duncan, D. B. *Biometrics* **1955**, 11, 1-42.
17) Agarwal, R.; Katiyar, S. K.; Zaidi, S. I. A.; Mukhtar, M. *Cancer Res.* **1992**, 52, 3582-3588.
18) Fujita, Y.; Yamane, T.; Tanaka, M.; Kuwata, K.; Okuzumi, J.; Takahashi, T; Okuda, T. *Jpn. J. Cancer Res.* **1989**, 80, 503-505.
19) Gao, Y. T.; McLaughlin, J. K.; Blot, W. J. *J. Natl. Cancer Inst.* **1994**, 86, 855-858.
20) Mitsui, T.; Yamada, K.; Yamashita, K.; Matsuo, N.; Okuda, A.; Kimura, G.; Sugano, M. *Intl. J. Oncol.* **1995**, 6, 377-383.
21) Yamada, K.; Watanabe, T.; Kaku, S.; Hassan, N.; Sugano, M. *Food Sci. Technol. Intl.* 1997, in press.
22) Beaven, M. A.; Rogers, J.; Moore, J. P.; Hesketh, T. R.; Smith, G.A.; Metcalfe, J.C. *J. Biol. Chem.* **1984**, 259, 7129-7136.
23) Lindau, M.; Gompert, B. D. *Biochim. Biophys. Acta* **1991**, 1071, 429-471.
24) Robinson, C.; Holgate, S. T. *J. Allergy Clin. Immunol.* **1985**, 76, 140-144.
25) Dahlen, S. E. , Kumlin, M.; Glanstrom, E.; Hedqvist, P. *Respiration* **1986**, 50, 22-29.
26) Kohiyama, N.; Nagata, T.; Fujimoto, S.; Sekiya, K. *Biosci. Biotech. Biochem.* **1997**, 61, 343-350.
27) Yamada, K.; Shirahata, S.; Murakami, H.; Nishiyama, K.; Shinohara, K.; Omura, H. *Agric. Biol. Chem.* **1985**, 49, 1423-1428.
28) Bauman, J.; von Bruchhausen, F.; Wurm, G. *Prostaglandins* **1980**, 20, 627-639.
29) Afanas'ev, I. B.; Dorozhko, A. I.; Brodskii, A. V.; Kostyuk, V. A.; Potapovitch, A. I. *Biochem. Pharmacol.* **1989**, 38, 1763-1769.
30) Husain, S. R.; Cillard, J.; Cillard, P. *Phytochem.* **1987**, 26, 2489-2491.
31) Morel, I.; Lescoat, G.; Cogrel, P.; Sergent, O.; Pasdeloup, N.; Brissot, P.; Cillard, P.; Cillard, J. *Biochem. Pharmacol.* **1993**, 45, 13-19.
32) Robak, J.; Gryglewski, R. J. *Biochem. Pharmacol.* **1988**, 37, 837-841.
33) Torel, J.; Cillard, J,; Cillard, P. *Phytochem.* **1986**, 25, 383-385.
34) Bors, W.; Heller, W.; Michel, C.; Saran, M. *Methods in Enzymol.* **1990**, 186, pp.343-355.
35) Folch, J.; Lees, M.; Sloane-Stanley, G. H. *J. Biol. Chem.* **1957**, 226, 497-506.
36) Ikeda, I.; Tomari, Y.; Sugano, M. *J. Nutr.* **1989**, 119, 1383-1387.
37) Das, N. P. Biochem. Pharmacol. **1971**, 20, 3435-3445.
38) Gu, J-Y.; Nonaka, M.; Yamada, K.; Yoshimura, K.; Takasugi, M., Ito, Y.; Sugano, M. *Biosci. Biotech. Biochem.* **1994**, 58, 1855-1858.

Chapter 21

Tea Catechin (EGCG) and Its Metabolites as Bioantioxidants

I. Tomita, M. Sano, K. Sasaki, and T. Miyase

School of Pharmaceutical Sciences, University of Shizuoka, 52-1 Yada, Shizuoka 422, Japan

After a single administration of 100mg (-)-epigallocatechin gallate (EGCG)/kg body weight orally to SD rats, its concentration changes in the plasma and bile were investigated using HPLC with electrochemical detector (ECD). The highest concentration in plasma was observed one hour after administration. Not only free, but also conjugated forms of EGCG (e.g. glucuronide and sulfate) were detected in both biological fluids. When EGCG was incubated with rat plasma or bile at 37 degrees C, most of it disappeared after 30 min, while about 20% of the antioxidative activities remained in both fluids. Two new dimerized products of EGCG, P-1 and P-2, were isolated from the incubation mixture with bile. They were identified as theasinensin A (P-1), and a new product (P-2). Though the yield of these 2 products from EGCG was small, their antioxidative activities comparable to that of EGCG, should be worth consideration.

It is known that (-)-epigallocatechin gallate(EGCG) is the major tea catechin in green tea leaves, and it exerts powerful antioxidative activities *in vitro(1,2)*, as well as *in vivo* experimental systems(*3-5*). We have shown that the oral administration of EGCG (50mg/kg) to rats once per day, for 7 days, inhibited *t*-butyl hydroperoxide (BHP) induced peroxidation in the liver and kidney slices up to 32% and 25%, respectively(*4*). We have also reported that feeding rats with 1% polyphenon 100 (catechin mixture of (+)-gallocatechin 1.4%, (-)-epicatechin 5.8%, (-)-epicatechin gallate 12.5%, (-)-epigallocatechin 17.6% and EGCG 53.9% supplied by Mitsui Norin Co.Ltd.) for 18 months, significantly suppressed the increase of TBARS level in plasma(*6*). It also suppressed the increase of triglyceride, total cholesterol and phospholipid in the plasma. These results indicate that catechins exert their activity *in vivo* after they were absorbed and distributed in rat tissue and organs. In relation to the activity of EGCG *in vivo,* its absorption, distribution and metabolic fate are of recent concern among investigators.

Unno et al. (*7,8*), and Okushio et al. (*9*), proved that EGCG was absorbed from rat or human intestinal tracts and it was present as such in plasma using GC-MS and LC-MS analysis. The highest concentration of EGCG in rat or human plasma was detected 1 or 2 hours after the administration and decreased quickly thereafter. Lee et al.(*10*), reported that catechins were present in their conjugated forms in the plasma and urine of humans who ingested 1.2g of decaffeinated green tea, containing 235mg catechins in total. The analysis was carried out by HPLC with the coulochem electrode array detection system. *In vivo* antioxidant effects of tea in men who ingested 300 ml of tea extract, prepared from 6g green tea leaves was recently reported by Serafini et al.(*5*), Maximum plasma radical trapping activity for peroxyl radicals was attained 30 min after the dosage of green tea.

In the present study, we have investigated the time course of the levels of free EGCG and the conjugated forms (glucuronide and sulfate) in the plasma and bile, after the oral administration of EGCG to free-moving rats with catheters. We also tried to identify the chemical structures of two compounds produced during the incubation of EGCG with bile.

Materials and Methods

Chemicals: EGCG was prepared from green tea leaves, which was processed at Shizuoka Tea Experimental Station. Its purity was confirmed to be *ca.* 98% by ^{1}H-NMR. 3,7-Dimethyl amino 10-*N*-methyl carbamoyl phenothiazin (MCDP) for antioxidative assay was obtained from Kyowa Medex Co. Ltd.(Tokyo, Japan). Beta-Glucuronidase (G-7896) and sulfatase (S-9754) were purchased from Sigma Chemical Co. (St.Louis. MO, USA). All other chemicals used were of reagent grade quality.

Animals: Male Sprague-Dawley rats (8 weeks old) were obtained from Japan SLC Inc. (Hamamatsu, Japan). Rats underwent both venous, and biliary catheterization to allow free moving, and enteral feeding under unanesthetized conditions. Blood and bile samples from the above rats were collected time-dependently for a period of 24 hours. The plasma was obtained by centrifuging the blood at 1500xg for 10 min. These samples were mixed with 1/5 volume of ascorbic acid-EDTA solution(*10*) and stored at –80 degrees C until analysis. Blood and bile used for the *in vitro* metabolic study of EGCG were collected from rats under EGCG anesthesia.

Extraction of EGCG and its dimerization products from plasma and bile: An aliquot of the sample was mixed with 1/4 volume of 0.1H HCl. The mixture was extracted 2 times with the same volume of ethyl acetate . The combined extract was evaporated and dried under N_2 gas. The residue was dissolved in 25% acetonitrile aqueous solution and used for the analysis of polyphenol and for the determination of antioxidant activities.

Analysis and assay for antioxidant potency. The analysis of EGCG and the dimerization products was performed by HPLC with ECD(*11*). Chemical structures of the dimerization products were elucidated on the basis of spectroscopic evidence such as UV, NMR, FAB-MS, SI-MS and CD. Incubation of the plasma and bile with beta-glucuronidase or sulfatase was carried out in the presence of ascorbic acid-EDTA by the method of Lee et al(*10*). Antioxidant activities (AOA) were evaluated using MCDP and radical inducers (*t*-butyl hydroperoxide, H_2O_2), or a thiobarbituric acid (TBA) assay with rat brain homogenates by the method of Sano, et al.(*12*) and Stocks, et al.(*13*), respectively.

Results and Discussion

Time course of the contents of EGCG and its conjugated forms in plasma and bile. After EGCG(100mg/kg-body weight) was administered orally to SD rats, portal blood and bile were periodically collected through free moving cannulation system. As shown in Figure 1, the maximum EGCG level was found 60 and 90 min after administration in plasma and bile, respectively. The level of EGCG in plasma decreased sharply after 1 hour, which is in accord with the results of Unno et al (7). On the contrary , the level of EGCG in bile decreased slowly with another peak in 5 hours, which might be a reflection of reabsorption through the intestinal tract. When the plasma or bile was treated with beta-glucronidase (500 units), or sulfatase (40 units), at 37 degrees C for 45 min in the presence of vitamin C and EDTA, their level changed as shown in Figure 2. The portion of glucronide to that of sulfate, and free form in plasma, increased after 120 min, while it was almost unchanged in bile. Their ratio was found to be 1:1:1, for glucronide, sulfate and free form.

Time course of antioxidative activity vs. metabolic change of EGCG in plasma and bile. When EGCG (100 ug/ml) was incubated with rat plasma or bile at 37 degrees C for 60 min, its content decreased sharply. Almost no EGCG was found after 45 min., as shown in Figure 3. The antioxidative activities (AOA) with BHP or H_2O_2, however, decreased rather moderately, and the activities remained in both plasma and bile after 45 min (see Figure 3). It was also found that antioxidative activities evaluated as TBARS with rat brain homogenate-autoxidation system(*13*) remained after 45 min (see also Figure 3). We, then, examined whether EGCG might be converted into other compounds with antioxidative activities in these fluids.

Isolation and identification of 2 dimeric products from EGCG. Plasma or bile, to which EGCG (100 ug/ml) was added, and incubated at 37 degrees C for 30 min, was analyzed by HPLC. It was found that 2 new peaks (P-1 and P-2) appeared, different from that of EGCG (Figure 4). The elucidation of the chemical structure of P-1 and P-2 were conducted using the products isolated from the reaction mixture of EGCG with bile. P-1 isolated as an amorphous powder has a molecular formula of

$C_{44}H_{34}O_{22}$ and $[alpha]_D^{24}$ was $-187.5\square$. The chemical structure was identical to that of theasinensin A(14) as shown in Figure 5. P-2 was a new compound, and afforded tridecaacetate by treatment with as amorphous powder. The molecular formula was determined as $C_{43}H_{32}O_{21}$ on the basis of secondary ion mass spectroscopy (SI-MS)

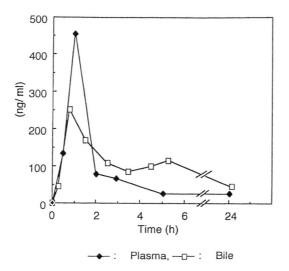

Figure 1. EGCG levels in rat plasma and bile following the oral administration of EGCG.
Values are average of duplicate determinations from two rats.

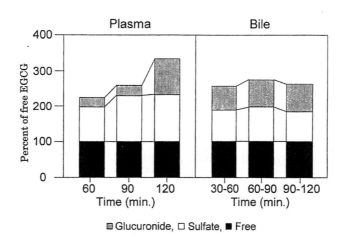

Figure 2. Free and conjugated EGCG in rat plasma and bile.

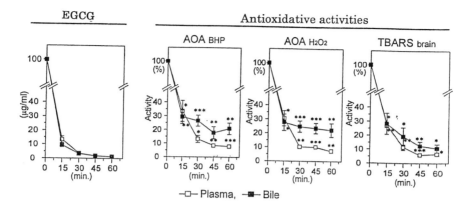

Figure 3. Fluctuation of antioxidative activities of EGCG in rat plasma and bile.
Values are mean±SE (plasma:n=4, bile:n=5). Significantly different from antioxidative activity of EGCG (*P<0.05, **:P<0.01, ***P<0.001).

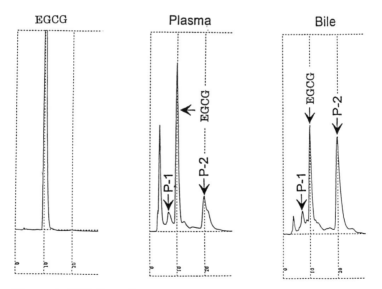

Figure 4. HPLC profiles of EGCG and dimerization products of EGCG (P-1,P-2).

and $[\text{alpha}]_D^{24}$ was $-189.0\square$. The chemical structure was elucidated from the analytical data of the ^{1}H-NMR, ^{13}C-NMR, difference nuclear Overhauser effect, and heteronuclear multiple bond connectivity spectra (see Figure 5).

Antioxidative activities of P-1, P-2 and their concentration changes in rat plasma and bile. It was suggested that both P-1 and P-2 would show antioxidative activity from the metabolic study of EGCG mentioned above. Antioxidative activity (AOA) was evaluated using BHP, or H_2O_2, and was designated as AOA_{BHP} and AOA_{H2O2}, respectively. AOA assay using rat brain homogenates(TBARS-brain) was also carried out. As shown in Table I, 50% inhibitory concentrations (IC_{50} uM) of both P-1 and P-2 were not much different from that of EGCG. It must be noted that IC_{50} for TBARS-brain of P-1 and P-2 were even smaller than that of EGCG.

Table I.   Antioxidative Activities of EGCG and its Dimerization Products (P-1, P-2)

Antioxidants	AOA_{BHP}	AOA_{H2O2}	TBARS-brain
		IC_{50} ($\square$ M)	
EGCG	9.6	11.6	0.20
P-1	10.2	12.1	0.12
P-2	13.1	14.2	0.19

Figure 6 shows the time course of the concentration of P-1 and P-2, when EGCG (100 ug/ml) was incubated with rat plasma or bile at 37 degrees C for 60min. With plasma, both P-1 and P-2 appeared shortly after the incubation of EGCG, and the maximum yields of P-1 and P-2 were 4.5% and 3.7% in respect to the amount of EGCG added. With bile, the formation of P-2 was much greater than that of P-1, and their yields in 15 min were 2.2% and 13.2 %, respectively. Though the pharmacological significance of P-1 and P-2 in the metabolism of EGCG should await further examination, their strong antioxidative activities should at least be fully taken into consideration. In discussing the antioxidative potency of tissues and organs of rats given EGCG, all of the activities might not necessarily be due to intact EGCG itself.

Acknowledgment

This work was supported by grant from Program for the Promotion of Basic Research Activities for Innovative Biosciences (Tokyo, Japan).

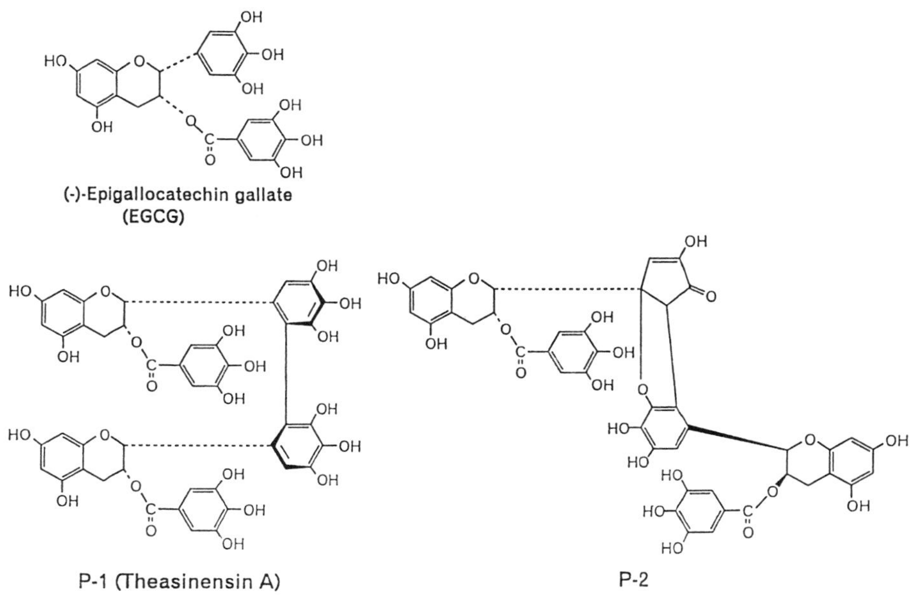

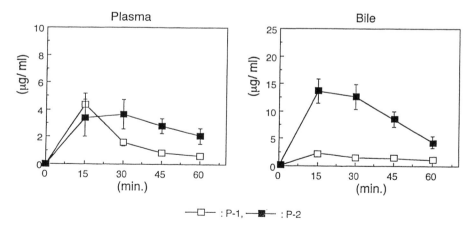

Figure 5. **Structures of EGCG and its dimers, P-1 and P-2.**

Figure 6. **Formation of dimerization products of EGCG in rat plasma and bile.**
Values are mean±SE (plasma:n=4, bile:n=5)

216

Literature Cited

1. Matsuzaki, T.; Hara,Y. *Nippon Nogeikagaku Kaishi* **1985**, *59*, 129-134.
2. Okuda,T.; Kimura,Y.; Yoshida,T.; Hatano,T.; Okuda,H.; Arichi,S. *Chem. Pharm. Bull.* **1983**, *31*,1625-1631.
3. Tomita, I.; Sano, M.; Watanabe, J.; Miura, S.; Tomita, I.; Yoshino, K.; Nakano, M. In *"Oxidative Stress and Aging"*; Cutler, R.G.; Packer, L.; Bertram, J.; Mori, A. Eds.; Birkhäuser Verlag : Basel, **1995**; pp.355-365.
4. Sano, M.; Takahashi, Y.; Yoshino, K.; Shimoi, K.; Nakamura, Y.; Tomita, I.; Oguni, I.; Konomoto, H. *Biol.Pharm.Bull.* **1995**, *18*, 1006-1008.
5. Serafini, M.; Ghiselli, A.; Ferro-Luzzi, A. *Eur. J. Clin. Nutr.* **1996**, *50*, 28-32.
6. Yoshino, K.; Tomita, I.; Sano, M.; Oguni, I.; Hara,Y.; Nakano, M. *Age* **1994**, *17*, 79-85.
7. Unno, T.; Takeo, T. *Biosci. Biotech. Biochem.* **1995**, *59*, 1558-1559.
8. Unno, T.; Kondo, K.; Itakura H.; Takeo, T. *Biosci. Biotech. Biochem.* **1996**, *60*, 2066-2068.
9. Okushio,K.; Matsumoto,N.; Kohri,T.; Suzuki,M.; Nanjo,F.; Hara,Y. *Biol.Pharm. Bull.* **1996**, *19*, 326-329.
10. Lee,M.J.; Wang,Z-Y.; Li, H.; Chen,L.; Sun,Y.; Gobbo,S.; Balentine,D.A.; Yang, C.S. *Cancer Epidemiology Biomakers & Prevention* **1995**, *4*, 393-399 .
11. Umegaki,K.; Esashi,T.; Tezuka,M.; Ono,A.; Sano,M.; Tomita,I. *J.Food Hygienic Soc. Jpn.* **1996**, *37*, 77-82 .
12. Sano,M.; Komatsu,T.; Tomita,I. In "VITAMIN KENKYU NO SHINPO", Kyoritsu Publishing Co., Ltd.: Tokyo, **1992**; *Vol.2*; pp.169-173.
13. Stocks,J.; Gutteridge,J.M.C.; Sharp,R.J.; Dormandy,T.L. *Clin. Sci. Mol. Med.* **1974**, *47*, 215-222.
14. Nonaka,G.; Kawahara,O.; Nishioka,I. *Chem.Pharm.Bull.* **1983**, *31*, 3906-3914.

Chapter 22

Effects of Tea Polyphenols Against *Helicobacter pylori*

M. Yamada[1], B. Murohisa[1], M. Kitagawa[1], Y. Takehira[1], K. Tamakoshi[1], N. Mizushima[1], T. Nakamura[1], K. Hirasawa[1], T. Horiuchi[1], I. Oguni, N. Harada[2], and Y. Hara[3]

[1]Gastrointestinal Division, Hamamatsu Medical Center, Hamamatsu, Japan
[2]Department of Food and Nutrition Sciences, University of Shizuoka, Hamamatsu College, Hamamatsu, Japan
[3] Food Research Laboratories, Mitsui Norin, Fujieda, Japan

Helicobacter pylori (H. pylori) is known to be associated with gastritis, peptic ulcer, and gastric carcinoma. We examined polyphenolic components of green tea such as epicatechin (EC), epigallocatechin (EGC), epicatechin gallate (ECg), epigallocatechin gallate (EGCg), for their antibacterial activity against H. pylori. Minimum inhibitory concentration (MIC) of each tea polyphenolic component against 23 strains of H. pylori was determined using agar dilution method. MIC50 were >200, 200, 50, 50 μg/mL for EC, EGC, ECg, EGCg, respectively. In addition, thirty-four patients who proved to be H. pylori infected were received green tea extracts for 1 month. Eradication of H. pylori was documented in 6 patients. These findings suggest that tea polyphenols have an antibacterial activity against H. pylori.

Helicobacter pylori (H. pylori) is a spiral-shaped, Gram-negative, urease-producing microaerophilic bacillus that has a strong affinity for gastric epithelium (*1*). It is estimated that more than half of the world's population is infected with this bacterium (*2*). During the last decade, H. pylori has been recognized as an agent responsible for chronic inflammation of the gastric mucosa. The infections are chronic, with little tendency to spontaneous cure, and are accompanied by persistent gastritis in nearly all infected individuals. H. pylori gastritis may produce gastric atrophy, a risk factor for gastric carcinoma, and intestinal metaplasia of the gastric mucosa, a potentially premalignant lesion. The natural history of the association between H. pylori and gastric cancer appears to involve acquisition of this organism in childhood followed by chronic gastritis for more than three decades, with progression to atrophic gastritis, intestinal metaplasia, dysplasia and finally to gastric cancer (3). This raises the exciting possibility that eradication of H. pylori infection could markedly reduce the incidence of this common and serious condition.

Tea extracts show bactericidal activity against various bacteria when added into the bacterial culture medium (4, 5). Polyphenol is one of the components responsible for this activity of tea extracts. The aim of the study was to investigate the effect of polyphenolic components of green tea such as epicatechin (EC), epigallocatechin (EGC), epicatechin gallate (ECg), and epigallocatechin gallate (EGCg) on the anti-H. pylori activity.

Materials and Methods

EC, EGC, ECg, EGCg were prepared from green tea as described previously (6). The molecular structures are illustrated in Figure 1. Bacterial susceptibility to catechins was tested by determining the minimal growth inhibitory concentration (MIC) for 3 standard strains (ATCC 43526, ATCC 43579, ATCC 43629) and 20 clinical isolates of H. pylori using agar dilution method. All strains were tested on Brucella HK agar supplemented with 10 % horse blood. Aliquots of H. pylori culture were transferred to the wells containing different concentrations of catechins and incubated at 35 degrees C in microaerobic atmosphere (5 % O_2, 10 % CO_2 for 72 h).

We next determined whether green tea extracts eradicate H. pylori in humans. Thirty-four patients with gastric ulcer, with duodenal ulcer, and with chronic gastritis who proved to be H. pylori infected were received capsules of tea extracts 700 mg/day orally for 1 month. Table 1 shows their components. The mean age of the patients was 54.2 years (range, 26 to 70 years); there were 22 males and 12 females. In each case, infection was confirmed when serological test (serum anti-H. pylori IgG antibody) and specific culture were positive. No patient received specific treatment for H. pylori before and during testing. A ^{13}C-urea breath test was performed to determine the H. pylori status of the patient before and after treatment. The breath test is a simple, non-invasive, and reliable test that reflects of ^{13}C-labeled urea by H. pylori urease (7-11). Following fasting for at least 6 h, patients were given a solution of 0.1 N citric acid to delay gastric emptying. After a further 5 min, patients were administered 150 mg of ^{13}C-urea dissolved in tap water. Duplicate breath samples were collected before and 15 min after administration of the ^{13}C-urea and were analyzed for the $^{13}C/^{12}C$ ratio in the CO_2 by isotope mass spectrometry. An increase of 6 per million in the $^{13}CO_2$ 15 min after ingestion of ^{13}C-urea compared with baseline measurement was considered positive for H. pylori. The ^{13}C-urea breath test was repeated in the patients whose clearance of H. pylori was defined as a negative breath test at the cessation of treatment. Recrudescence of H. pylori infection usually occurs within 4 weeks after cessation of treatment if the organism is not completely eliminated; if a patient is still H. pylori negative 4 weeks after treatment he or she has been presumed likely to remain so (1, 12-15). Therefore eradication of H. pylori was assessed as a negative breath

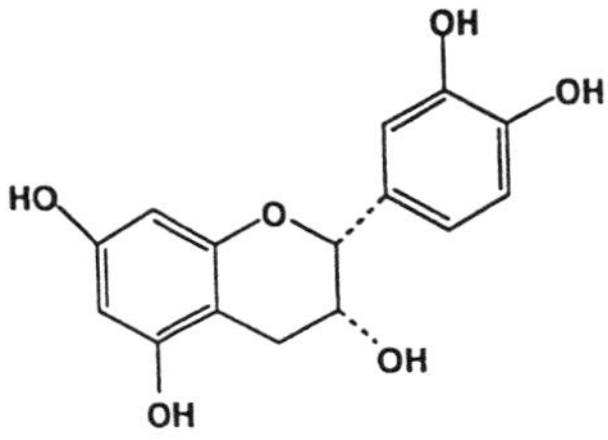

Fig. 1. Molecular structures of green tea polyphenols.

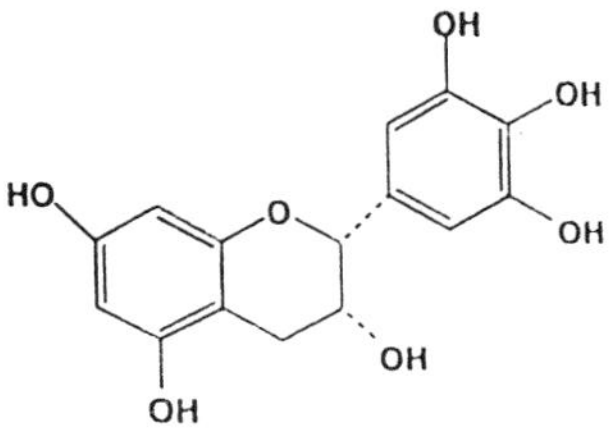

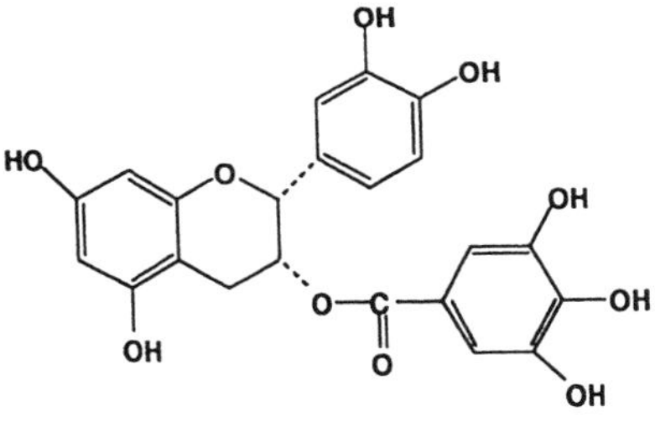

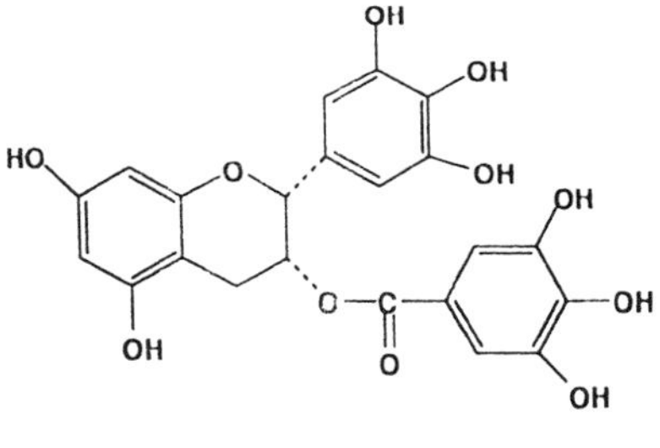

test at one month after the end of treatment. Changes in biochemical blood analysis and urinalysis were assessed before and 1 month after the treatment.

Table 1. Composition of green tea polyphenols in catechin capsule

Green tea catechins	%
Epigallocatechin (EGC)	21.0
Epicatechin (EC)	7.3
Epigallocatechin gallate (EGCg)	29.2
Epicatechin gallate (ECg)	7.9
Total	65.4

Results

MIC_{50} of tea polyphenols against clarithromycin sensitive strains of H. pylori were >200 (25 − > 200), 200 (3.13 − >200), 50 (6.25 − 100), 50 (3.13 − 100) μg/mL for EC, EGC, ECg, EGCg, respectively. MIC_{50} values against clarithromycin resistant strains were same without that of EC. Either ECg or EGCg was the most potent agent tested against H. pylori. These results were similar to those against standard strains (Table 2).

In the second part of the study, all 34 patients completed the course of treatment, and no adverse event was recorded in biochemical analysis and urinalysis. More than half of 34 patients were found to be decreased in the delta (Δ) 0/00 values one month after the treatment, and eradication of H. pylori was documented in 6 patients one month after the completion of treatment (Figure 2).

Discussion

Geographical studies of gastric cancer incidence and H. pylori prevalence reveal striking parallels between the two. Gastric cancer, like H. pylori infection, has foci of high incidence, usually in populations of low socioeconomic class (*16,17*). In 1994, the International Agency for Cancer Research recognized a cause-and-effect relationship between H. pylori and gastric adenocarcinoma and classified H. pylori as a group I (definite) carcinogen (*18*). Unfortunately, there is no simple and entirely safe means of eradicating H. pylori infection. The "gold standard" thus far has been triple therapy with bismuth compounds, metronidazole, and tetracycline (*12, 13*).

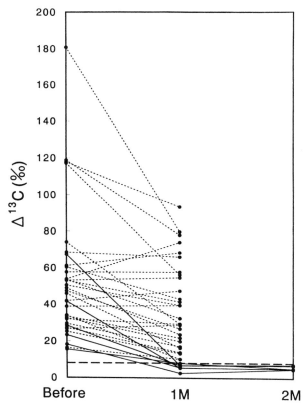

Fig. 2. Changes in H. pylori activity before and after catechin administration to H. pylori-infected patients. Catechin (700 mg/day) was orally administered to H. pylori-infected 34 patients for one month. 13C-urea breath test was performed to determine the H. pylori status of the patients before and after the treatment (1M, 2M). Eradication was defined as a negative breath test at one month after the cessation of treatment (2M). Urea breath test consisted of a baseline breath sample and a breath sample 15 minutes after the administration of 150 mg of 13C-urea dissolved in tap water. Breath samples were measured by mass spectrometry. Values were expressed as excess delta (Δ) per mil units, which were the ratio of 13C to 12C in the sample compared with a standard, multiplied by 1000, minus the baseline value.

•---•; patients with cure of H. pylori infection, ———; patients without cure of H. pylori infection

Table 2. Minimum inhibitory concentration of catechins for standard strains and clinical isolates of H. pylori (µg/mL)

	EC	EGC	ECg	EGCg
ATCC 43526	> 200	200	50	50
ATCC 43629	> 200	> 200	50	50
ATCC 43579	> 200	200	50	50
CAM (-) (n = 8)	200*	200*	50*	50*
CAM (+) (n = 12)	> 200*	200*	50*	50*

Abb: ATCC, american type culture collection (standard strain); CAM (-), clarithromycin resistant strain; CAM (+), clarithromycin sensitive strain; EC, Epicatechin; EGC, Epigallocatechin; ECg, Epicatechin gallate; EGCg, Epigallocatechin gallate.
*MIC_{50} values were expressed.

None of the antibiotic treatments are particularly easy to take and approximately 30 % of patients develop some sort of side effect, usually nausea and diarrhea or headache. Occasionally the treatment can result very severe diarrhea and the development of pseudomembranous colitis (*19*). Moreover, there is the tendency of the antibiotic treatment to increase bacterial resistance within the general population (*20*). Tea leaves contain various kinds of polyphenols which are also called tea catechins. The principal catechins present in green tea are (-) - EC, (-) - EGC, (-) - ECg, and (-) - EGCg. Catechins with the presence of galloyl and gallic moieties of the catechin structure were, in particular, highly effective in promoting antibacterial activity against H. pylori in vitro. Although the mechanism of how these compounds damage H. pylori remains to be examined in the future, it is suggested that the catechins interact with the bacterial membrane and perturb the lipid bilayer and disrupting the barrier function (*21*). Our results suggest that green tea polyphenols have an antibacterial activity against H. pylori under the concentration of as in ordinarily brewed tea (1000 µg/mL) and these compounds may be useful as prophylactic and therapeutic agents against H. pylori infection.

Acknowledgment

This work has been supported in part by a grant from the Program for Promotion of Basic Research Activities for Innovative Biosciences.

Literature Cited

1. Marshall, B. J. Helicobacter pylori. *Am. J. Gastroenterol.* **1994,** *89,* S116–128.

2. Lee, A. The nature of Helicobacter pylori. *Scand. J. Gastroenterol.* **1996,** *31(Suppl 214),* 5–8.

3. Munoz, N.; Connelly, R. Time trends of intestinal and diffuse types of gastric cancer in the United States. *Int. J. Cancer* **1971,** *8,* 158–164.

4. Toda, M.; Okubo, S.; Hiyoshi, R.; Shimamura T. The bactericidal activity of tea and coffee. *Lett. Appl. Microbiol.* **1989,** *8,* 123–125.

5. Toda, M.; Okubo, S.; Ikigai, H.; Suzuki, T.; Suzuki, Y.; Shimamura, T. The protective activity of tea against infection by Vibrio cholerae O1. *J. Apply. Bacteriol.* **1991,** *70,* 109–112.

6. Okushio, K.; Matsumoto, N.; Kohri, T.; Suzuki, M.; Nanjo, F.; Hara, Y. Absorption of tea catechins into rat portal vein. *Biol. Pharm. Bull.* **1996,** *19,* 326–329.

7. Graham, D. Y.; Klein, P. D.; Evans, D. J. Jr; Evans, D. G.; Alpert, L. C.; Opekun, A. R.; Boutton, T. W. Campylobacter pylori detected non-invasively by the ^{13}C-urea breath test. *Lancet* **1987,** *1,* 1174–1177.

8. Dill, S.; Payne-James, J. J.; Misiewicz, J. J; Grimble, G. K.; McSwiggan, D.; Pathak, K.; Wood, A. J.; Scrimgeour, C. M.; Rennie, M. J. Evaluation of ^{13}C-urea breath test in the detection of Helicobacter pylori and in monitoring the effect of tripotassium dicitratobismuthate in non-ulcer dyspepsia. *Gut* **1990,** *31,* 1237–1241.

9. Logan, R. P.; Polson, R. J.; Misiewicz, J. J.; Rao, G.; Karim, N. Q.; Newell, D.; Johnson, P.; Wadsworth, J.; Walker, M. M.; Daron, J. I. I. Simplified single sample [13]Carbon urea breath test for Helicobacter pylori: comparison with histology, culture, and ELISA serology. *Gut 1991,* **32,** 1461–1464.

10. Klein, P. D.; Graham, D. Y. Minimum analysis requirements for the detection of Helicobacter pylori infection by the ^{13}C-urea breath test. *Am. J. Gastroenterol* **1993,** *88,* 1865–1869.

11. Braden, B.; Duan, L. P.; Caspary, W. F.; Lembcke, B. More convenient ^{13}C-urea breath test modifications still meet the criteria for valid diagnosis of Helicobacter pylori infection. *Z. Gastroenterol.* **1994,** *32,* 198–202.

12. Weil, J.; Bell, G. D.; Jones, P. H.; Gant, P.; Trowell, J. E.; Harrison, G. "Eradication" of Campylobacter pylori: Are we misled? *Lancet* **1988,** *2,* 1245.

13. Tytgat, G. N. J.; Noach, L.; Rauws, E. A. J. Helicobacter pylori. *Scand. J. Gastroenterol.* **1991,** *26(suppl 187),* 1–8.

14. Logan, R. P. H.; Dill, S.; Bauer, F. E.; Walker, M. M.; Hirschl, A. M.; Gummett, P. A.; Good, D.; Mossi, S. The European ^{13}C-urea breath test for the detection of Helicobacter pylori. *Eur. J. Gastroenterol. Hepatol.* **1991,** *3,* 915–921.

15. Bell, G. D. Clinical aspects of infection with Helicobacter pylori. *Commun. Dis. Rep. Rev.* **1993,** *3,* R59–62.

16. Perez-Perez, G. I.; Taylor, D. N.; Bodhidatta, L.; Wongsrichanalai, J.; Baze, W. B.; Dunn, B.E.; Echeverria, P. D.; Blaser, M. J. Seroprevalence of Helicobacter pylori infections in Thailand. *J. Infect. Dis.* **1990**, *161*, 1237–1241.

17. EUROGAST Study Group. An international association between Helicobacter pylori infection and gastric cancer. *Lancet* **1994**, *341*, 1359–1362.

18. International Agency for Research on Cancer. Schistosomes, liver flukes and Helicobacter pylori. IARC monographs on the evaluation of carcinogenic risks to humans. 1994; Vol 61, Lyon: IARC.

19. Thijs, J.C.; van Zwet, A. A.; Oey, H. B. Efficacy and side effects of a triple drug regimen for the eradication of Helicobacter pylori. *Scand. J. Gastroenterol* **1993**, *28*, 934–938.

20. Biasco, G.; Miglioli, M.; Barbara, L.; Corinaldesi, R.; DiFebo, G. Omeprazole, Helicobacter pylori, gastritis and duodenal ulcer. *Lancet* **1989**, *ii*, 1403.

21. Ikigai, H.; Nakae, T.; Hara, Y.; Shimamura T. Bactericidal catechins damage the lipid bilayer. *Biochim. Biophys. Acta.* **1993**, *1147*, 132–136.

Chapter 23

Modulation of Mitotic Signal Transduction by Curcumin and Tea Polyphenols and Their Implication for Cancer Chemoprevention

Jen-Kun Lin, Yu-Chih Liang, Yu-Li Lin, Yen-Chou Chen, and Shoei-Yn Lin Shiau

Laboratory for Cancer Research, Institutes of Biochemistry and Toxicology, College of Medicine, National Taiwan University, Taipei, Taiwan, Republic of China

It has been demonstrated that diets rich in fruits and vegetables are protective aganist cardiovascular diseases and certain forms of cancer. These protective effects have been attributed to the anti-oxidant present, including vitamin C, Carotenes and phytopolyphenols. The polyphenolic components of higher plants may act as anti-oxidants (sometimes may be as pro-oxidants), or as agents of other mechanisms, contributing to the anti-carcinogenic or cadioprotective action. Curcumin is a widely used dietary pigment (curry), and this polyphenolic compound has been demonstrated to be an inhibitor of tumor promotion in chemical carcinogenesis. Recent studies also indicate that tea polyphenols are active in inhibiting the processes of carcinogenesis induced by various carcinogens. Investigations from this laboratory and others have indicated that modulation of mitotic signal transduction may attribute, in part, to the molecular mechanisms of these cancer chemopreventive agents.

Carcinogenesis is a multiple process comprising initiation, promotion and progression. Multistage models of carcinogenesis have been developed for various tissues and animal species, for mechanistic investigations (1, 2). These animal models have played important roles as in vivo test systems in the identification of exogenous and/or endogenous agents exerting a role in enhancement, or prevention, of various stages of carcinogenesis (2). It has been suggested that intervention of the carcinogenesis process by chemical means, namely cancer chemoprevention, can affect at any of the three stages. However, the promotion stage, because of its reversibility may be a particularly attractive candidate for chemical intervention. Tumor promotion has been described as the clonal expansion or proliferation of an initiated cell. Two distinct effects of tumor promoting chemicals have been demonstrated : direct mitogenic effects (proliferation), and the induction of cell lethality (apoptosis). It is apparent that cell proliferation, induced through mitogenic effects may occur directly by interaction of the chemical with a receptor. It may also occur by direct modification of gene expression, or indirectly by the modification of growth factors or hormonal stimulation/inhibition. Meanwhile, tumor promotion may also be induced through partial cytolethal means ; the induction of cell death by a chemical in the target tissue results in a compensatory increase in cell proliferation in the surrounding surviving cells (3).

A number of tumor promoters have been shown to produce free oxygen radicals in cells and tissues (Oxygen radicals have been demonstrated to participate in mouse skin tumor promotion) (4). The production of reactive oxygen species (ROS) either extracellularly or intrcellularly, by tumor promoting compounds, has been associated with the tumor promotion stage in several animal model systems (4,5). If the formation of ROS and the resulting damage to cellular proteins, lipids and/or nucleic acids are important in this stage of promotion, then chemopreventive agents that function through anti-oxidant mechanisms may be effective in preventing the transcription of the initiated cell to the neoplastic state.

Anti-oxidants are available from both natural and synthetic sources. Chinese herbal medicine such as tumeric and tea are two important plants containing strong antioxidants. Chinese green tea, in particular is composed of several catechin polyphenols. These tea polyphenols have anti-oxidant properties and have been demonstrated to be anticarcinogenic as well as antimutagenic (6). Tea polyphenols have demonstrated chemopreventive activities in animal cancer models in the colon and large intestine, duodenum, esophagus, forestomach, liver, lung, mammary gland and skin (7).

Curcumin is the major yellow pigment in turmeric and curry and is obtained from the rhizome of the plant curcuma longa. Curcumin had chemopreventive activity in mouse skin, colon (8) and rat mammary models (9).

Biochemical mechanisms of action of curcumin

Curcumin is a potent anti-inflammatory agent (10); it inhibits arachidonic acid metabolism by blocking both the lipooxygenase and cyclooxygenase pathways, and possibly phospholipase A2. Curcumin exhibits its strong antioxidant activity (11) by being an effective superoxide scavenger (12). It inhibits TPA-induced DNA synthesis, demonstrating an inhibitory effect on proliferation (8). It also modifies cytochrome p450, enhances glutathione-S-transferase activity, and may modify the metabolic activation and DNA binding of PAH carcinogens by this mechanism. Curcumin has also been shown to inhibit the activities of cellular kinases, including phosphorylase kinase (13), and EGF-R kinase (14). The short-term treatment of cells with curcumin inhibited the EGF receptor intrinsic kinase activity up to 90% in a dose- and time-dependent manner. It also inhibited the EGF-induced tyrosine phosphorylation of EGF receptors (14). The observed early effects of curcumin were mediated via cellular mechanisms, and preceded the period when inhibition of cell growth occurred. Furthermore, curcumin has been shown to inhibit the TPA-mediated induction of c-Jun/AP-1 (*15*), ornithine decarboxylase (*8*), and protein kinase C activity (*16*) in mouse NIH3T3 cells.

Inhibition of EGF-mediated tyrosine phosphorylation of EGF-R, by curcumin, was mediated by a reversible mechanism. In addition, curcumin also inhibited EGF-induced, but not bradykinin-induced calcium release. These findings demonstrate that curcumin is a potent inhibitor of a growth stimulatory pathway, the ligand-induced activation of EGF-R, and may potentially be useful in developing anti-proliferative strategies to control tumor cell growth (*17*).

Induction of apoptosis by curcumin

It has demonstrated that many antitumor agents are capable of inducing cell death by the processes of apoptosis. Curcumin could stop the growth of immortalized mouse embryo fibroblast NIH3T3 cells, human colon cancer cell HT-29, human kidney cancer cell 293 and human hepatoma HepG2 cells with various common features of apoptosis (*18*). The apoptosis changes were not detected in the primary culture of the mouse embryonic fibroblast C3H10T1/2, rat embryonic fibroblast and human foreskin fibroblast cells. Treatment of NIH3T3 cells with the PKC inhibitor staurosporine, the tyrosine inhibitor herbimycin A, and the arachidonic acid metabolism inhibitor quinacrine induces apoptotic cell death. These results suggest that in some immortalized and transformed cells, blocking the cellular signal transduction might trigger the induction of apoptotsis (*18*).

We also found that curcumin could induce apoptotic cell death in promyelocytic leukemia HL-60 cells, at concentration as low as 3.5 μg/ml. The apoptotic induction of curcumin appeared in a dose- and time-dependent manner (*19*). Flow cytometric analysis showed that the hypodiploid DNA peak of propidium iodide-stained nuclei appeared at 4 hr after 7 μg/ml curcumin treatment. The apoptotic induction activity of curcumin was not affected by cycloheximide, actinomycin D, EGTA, W7 (calmodulin inhibitor), sodium orthovanadate, or genistein. By contrast, an endonuclease inhibitor $ZnSO_4$ and proteinase inhibitor N-tosyl-L-lysine chloromethyl ketone (TLCK) could markedly abrogate apoptosis induced by curcumin, while TPA had a partial effect. The antioxidants, N-acetyl-L-cysteine, L-ascorbic acid, α-tocopherol, catalase, and superoxide dismutase, all effectively prevented curcumin-induced apoptosis. These result suggested that curcumin-induced cell death was mediated by ROS. Immunoblot analysis showed that the level of the antiapoptotic protein Bcl-2 was decreased to 30% after 6 hr treatment with curcumin, and was subsequently reduced to 20% by a further 6 hr treatment. Furthermore, overexpression of bcl-2 in HL-60 cells resulted in a delay of curcumin-treated cells entering into apoptosis, suggesting that bcl-2 plays a crucial role in the early stage of curcumin-triggered apoptotic cell death in this cell line (*19*).

Apoptosis of HepG2 cells, triggered by curcumin and other agents, is characterized in an attempt to delineate the common apoptosis signaling pathway in human hepatoma cells. Several hallmarks of apoptosis, including DNA ladder, chromatin condensation, DNA fragmentation, and an apoptosis specific cleavage of 28S and 18S ribosomal RNA were observed after treatment with curcumin (*20*). Curcumin treatment, however, did not alter the expression levels of Bax proteins.

p53 protein accumulated slowly, and decreased abruptly after reaching the maximum. Conversely, c-myc protein decreased initially, and subsequently increased proceeding the onset of apoptosis (20). When COLO 205 colorectal carcinoma cells were treated with curcumin (60 µM), the appearance of apoptotic DNA ladders was delayed about 5h, and G1 arrest was detected (21). The reduction of p53 gene expression was accompanied by the induction of the Hsp70 gene expression in the curcumin treated cells. The intracellular Ca^{+2} concentration was depleted by curcumin in a dose-dependent manner. These findings suggest that curcumin may induce the expression of Hsp 70 gene through the initial depletion of intracellular Ca^{+2}, followed by the suppression of p53 gene function in the target cells (21).

Biochemical mechanisms of action of tea polyphenols

Tea has been considered as a crude medicine in China for more than 4,000 years. Different kinds of pharmaceutical effects, such as protection of blood vessels, reduction of serum cholesterol levels, and prevention of arteriosclerosis were reported as an integrated effect (22). Green tea contains a large amount of polyphenols, most of which are flavan-3-ols, commonly known as catechins. The major flavan-3-ols are (-)-epicatechin derivatives including (-)-epicatechin (EC), (-)-epigallocatechin (EGC), (-)-epicatechin-3-gallate (ECG) and (-)-epigallocatechin-3-gallate (EGCG).

The biochemical mechanisms of action of these tea polyphenols have been intensively investigated (23). The biochemical and pharmacological effects of tea polyphenols have been attributed to the competitive inhibition of cytochromes p450 involved in the bioactivation of various carcinogens, as well as to antioxidant properties as scavengers of ROS (24). Recent studies have shown that tea polyphenols cause increases in the activities of phase II detoxifying enzymes, such as glutathione-S-transferase, NAD(P)H quinone reductase, epoxide hydrolase, and UDP-glucuronosyl transferases (23).

It is known that the 5'-flanking regions of phase II genes contain several cis-acting regulatory elements, such as the antioxidant-responsive element (ARE)/electrophile-responsive element and xenobiotic-responsive element (XRE)/aromatic hydrocarbon-responsive element, which are through to mediate the induction of phase II enzymes by many drugs (25,26). It is suggested that the stimulation of mitogen-activated protein kinases (MAPKs) may be the potential signaling pathways utilized by tea polyphenols to activated ARE-dependent genes (27).

Suppression of extracellular signals and cell proliferation by tea extract and EGCG

We have demonstrated that the major and most potent component of tea polyphenol EGCG inhibited the growth of S-180 and A431 cells (28). The mechanism underlying these inhibitory effects of EGCG is not fully understood. For this reason, we tried to explore the growth inhibiting nature, and mechanism of action, of tea extract and EGCG on human epidermoid carcinoma A431 cells that express high levels of epidermal growth factor (EGF) receptors (*29*). The EGF receptor is a 170 kDa plasma membrane glycoprotein with an extracellular ligand-binding domain, a single transmembrane region, and an intracellular domain that exhibits intrinsic tyrosine kinase activity (*30*).

Various aqueous extracts from Green tea, Paochung tea, Oolong tea and Black tea, showed the significant inhibitory effects on the binding of ^{125}I-EGF to A431 cells (Figure 1). The degree of these inhibitions are dose-dependent. Furthermore, the inhibitions of tea extracts on the phosphorylation of the EGF receptor were demonstrated *in vivo*, in A431 cells (Figures 2&3). It seems that the inhibitions are dose-dependent, and Green tea extract is stronger as compared to other tea extracts. The major tea polyphenol in the Green tea extract has been shown to be EGCG. Therefore, we promptly researched the effect of EGCG on the phosphorylation of EGF-receptor (EGF-R).

In the experiments shown in Figure 4, we examined the effects of different concentrations of EGCG on the EGF-R autophosphorylation of A431 cells, in response to EGF (Figure 4A). A similar experiment was performed to see the effect of EGCG on the PDGF-R autophosphorylation of NIH3T3 cells in response to PDGF (Figure 4B). Western blotting, using an anti-tyrosine antibody, showed that EGF-R (or PDGF-R) was rapidly tyrosine phosphorylated upon EGF (or PDGF) stimulation. The results indicate that 1 μg/ml of EGCG is sufficient to inhibit the EGF-R kinase activity by 45% (Figure 4A), while 5 μg/ml of EGCG is able to inhibit the PDGF-R kinase activity by 53% (Figure 4B). There is no effect of EGCG treatment (30 min) on the level of total EGF-R protein (Figure 4A, bottom), or PDGF-R protein (data not shown). These results suggest that EGCG can suppress the extracellular signals through suppressing EGF-R (or PDGF-R) autophosphorylation upon EGF (or PDGF) stimulation (*31*).

In order to determine the amino acid target for phosphorylation, A431 cells were labeled with [^{32}p]-orthophosphate, then pretreated with EGCG before EGF (*31*). EGF-R was immunoprecipitated from total cell lysates, and analyzed by SDS-PAGE, followed by autoradiography. There is an increase in the level of total phosphorylation of the ^{32}p-EGF-R by EGF (20 ng/ml, Figure 5A, lane 2), and it can be inhibited by EGCG (5 μg/ml, Figure 5A, lane 3). Phosphoamino acid analysis of the EGF-R, by two-dimensional thin layer electrophoresis (TDTLE) showed that EGCG only decreased the phosphorylation of the receptor protein on the tyrosine (Y) residue (Figure 5B), but not on serine (S) and threonine (T) residues (*31*).

Inhibition of receptor tyrosine kinase activity by EGCG in vitro

We examined the effect of EGCG on the activities of six different protein kinases. The results indicated that EGCG inhibited all of the kinases examined, but to different extents (Table I). The inhibitory activity of EGCG is more effective in receptor-type protein tyrosine kinases (EGF-R, PDGF-R, and FGF-R, IC_{50}=0.5-1.0 μg/ml), than non-receptor-type protein tyrosine kinase (pp60$^{v\text{-src}}$, $IC_{50} > 10$ μg/ml). In receptor-type protein kinases (PTKs), EGCG shows more selective inhibition of EGF-R ($IC_{50} = 0.5$ μg/ml) than other receptor-type PTKs (*31*). By contrast, EGCG

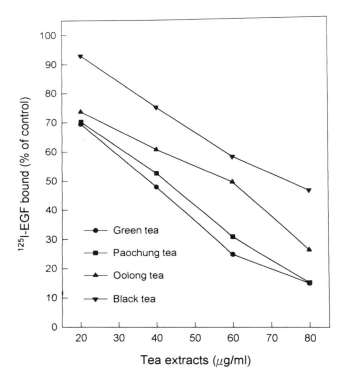

Effects of various tea extracts on ^{125}I-EGF binding to A431 cells.

Figure 1. Effect of various tea extracts on ^{125}I-EGF binding to A431 cells. Cells were serum-starved for 6h and then treated with various concentrations of tea extracts. After 30 min, ^{125}I-EGF were added for 1h. After the incubation, the cells were washed and solubilized with 1 ml of 1.5 N NaOH at RT for 1h. The solubilized lysates were transferred to plastic tubes for counting on a γ-spectrometer.

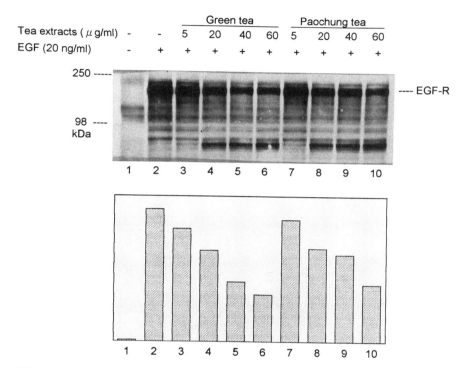

Figure 2. Effect of Green tea and Paochung tea extracts on tyrosine phosphorylation in A431 cells induced by EGF. Serum-starved were exposed to various concentrations of tea extracts for 30 min and then to 20 ng/ml of EGF for 10 min. Total cellular proteins (50 µg) were separated on SDS-PAGE (10% polyacrylamide) and blotted with anti-phosphotyrosine antibody. Immuno-complexes were detected by horseradish peroxidase second antibody and then by ECL kits. The position of the 170-kDa phosphotyrosine protein is indicated as EGF-R at the right (top). The intensity of EGF-R phosphorylation bands were quantified by densitometry (IS-1000 Digital Imaging System) (bottom).

232

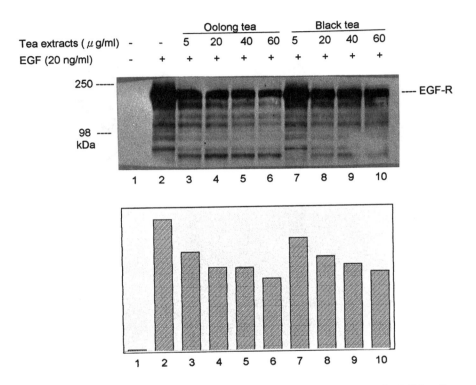

Effects of Oolong tea and Black tea on tyrosine phosphorylation in A431 cells induced by EGF .

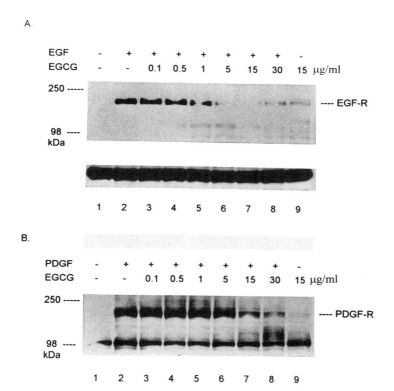

Figure 4. Effect of EGCG on tyrosine phosphorylation in A431 (A) or NIH3T3 (B) cells induced by EGF (A) or PDGF (B). Serum-starved cells were exposed to various concentrations of EGCG for 30 min (lanes, 3-9) and then to 20 ng/ml of EGF (or 10 ng/ml of PDGF) (lanes, 2-9). Expression of tyrosine phosphorylated proteins was measured as described (31). Expression of EGF-R was analyzed by immunoblotting using first anti-EGF-R mAb.

(Reproduced with permission from reference 31. Copyright 1997 Wiley.)

A.

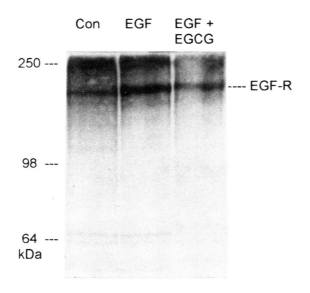

B.

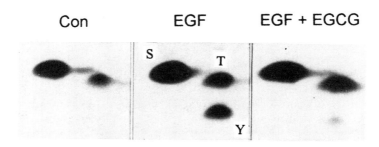

Figure 5. Analysis of EGF-induced phosphorylation of EGF-R by EGCG. A: Cells were labeled with [^{32}p]orthophosphate for 4h, and incubated with or without EGCG (5 μg/ml) for 30 min then EGF (20 ng/ml) for 10 min. ^{32}p-labeled EGF-R were immunoprecipitated with anti-EGF-R mAb and analyzed by autoradiography. B: Two-dimensional thin-layer electrophoresis (TDTLE) pattern of amino acids after hydrolysis of ^{32}p-EGF-R excised from Immobilon-P membrane. S, phosphoserine; T, phosphothreonine; Y, phosphotyrosine.

(Reproduced with permission from reference 31. Copyright 1997 Wiley.)

scarcely inhibits serine- and threonine-specific protein kinases such as protein kinases A and C at 20 µg/ml (Table I). Thus inhibitory activity of EGCG is highly specific for receptor-type PTKs, especially EGF-R. This observation prompted us to investigate the effects of EGCG on the DNA synthesis in A431 cells. Table II shows that EGCG alone has no significant effect on cell survival at 15 µg/ml, but can inhibit the [³H]-thymidine incorporation into DNA in EGF-treated cells (31). The activity of ERK 1 and ERK 2 was inhibited by EGCG (Figure 6).

Table I. Effect of EGCG on the activities of protein kinases in vitro

Protein kinase	IC_{50} (µg/ml)[a]
EGF receptor	0.5
PDGF receptor	1
FGF receptor	1
pp60[v-src]	> 20
Protein kinase C	> 20
Protein kinase A	> 20

[a]The activity of protein kinase was measured as described (31).

Table II. Effect of EGCG on EGF-stimulated [³H]-thymidine incorporation into DNA in A431 cells.

Treatment	[³H] Thymidine incorporation (relative ratio to control)[a]
Control	1.00
EGF (40 ng/ml)	1.62 ± 0.02
EGCG (15 µg/ml)	0.92 ± 0.11
EGF + EGCG (1 µg/ml)	1.68 ± 0.06
EGF + EGCG (5 µg/ml)	1.42 ± 0.08
EGF + EGCG (15 µg/ml)	1.06 ± 0.04

[a]Each experiment was independently performed three times (31) and expressed as mean ± SE.

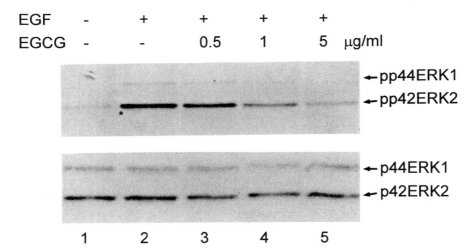

Figure 6. Effect of EGCG on the inhibition of MAP kinase (ERK1 and ERK2) in A431 cells. Serum-starved cells were exposed to various concentrations of EGCG for 30 min and then to 20 ng/ml of EGF for 10 min. Total cellular proteins (50 μg) were separated on SDS-PAGE (10% polyacrylamide) and blotted with anti-phospho-MAPK antibody (top) or anti-MAPK antibody (bottom). Immunocomplexes were detected by alkaline phosphatase second antibody and then by NBT/BCIP substrates.

Induction of nitric oxide synthase was blocked by EGCG

Nitric oxide (NO) plays an important role in inflammation, and multiple stages of carcinogenesis. We investigated the effect of various tea polyphenols on the induction of NO synthase (NOS), in thioglycollate-elicited and lipopolysaccharide (LPS)-activated peritoneal macrophages. Gallic acid, (-)-epigallocatechin and EGCG the major tea catechin, were found to inhibit inducible NOS (iNOS) protein in activated macrophage. EGCG, a potent antitumor agent with anti-inflammatory and antioxidant properties, inhibited NO generation, as measured by the amount of nitrite released into the culture medium (*32*).

An electrophoretic mobility shift assay indicated that EGCG blocked the activation of NFκB, a transcription factor necessary for iNOS induction. EGCG also blocked the disappearance of IκB from the cytosolic fraction. These results suggest that EGCG decreases the activity and protein levels of iNOS by reducing the expression of iNOS mRNA. The reduction could occur through prevention of the binding of NFκB to the iNOS promoter, thereby inhibiting the induction of iNOS transcription (*32*).

Concluding remarks

It has been demonstrated that the function of many growth signal transducing proteins is significantly dependent on their redox state. Such proteins include growth factor receptors, protein kinases, protein phosphatases, as well as a number of important transcription factors, including AP-1 and NFκB (*33*). It has been speculated that the adjustment of the redox states of individual signal-transduction proteins within cells is aprerequisite for their optimum functioning in the transmission of growth response (*33*). Such protein redox regulation could be achieved through the direct oxidative interaction of H_2O_2 (or O_2^-), or indirectly through changes in cellular levels of GSH and GSSG.

Both curcumin and EGCG are active anti-oxidants. As described above, these two compounds can act as ROS scavenger and inhibit the activities of protein kinases and growth factor receptor kinases. They can effectively suppress the activation of NFsymbol 107 \f "Symbol" \s 12κB. Therefore, it is very likely that both curcumin and EGCG (and other tea polyphenols) may function as cancer chemopreventive agents through modulation of the redox mitotic signal transduction. The molecular mechanisms of this modulation deserve further investigation.

Acknowledgments

This study was supported by the National Science Council NSC-86-2314-B002-042 and the National Health Research Institute Grant, DOH 86-HR-403.

Literature Cited

1. Hennings, H.; Glick, A. B.; Greenhalgh, D. A.; et al. *Proc. Soc. Exp. Biol.* **1993**, 202, 1-8.

2. Slaga, T. J.; Budunova, I. V.; Gimenez-Conti, I. B.; Aldaz, C. M. *J. Invest. Dermatol. Sym. Proc.* **1996**, 1, 151-156.
3. Klaunig, J. E. *Prev. Med.* **1992**, 21, 510-519.
4. Klein. Szanto, A. J. P. ;Slaga, T. J. *J. Invest. Dermatol.* **1982**, 79, 30-34.
5. Pitot, H.; Goldsworthy, T.; Campbell, H.; Poland, A. *Cancer Res.* **1980**, 40, 3616-3620.
6. Ruch, R. J.; Cheng, S.-J.; Klaunig, J. E. *Carcinogrenesis* **1989**, 10, 1003-1008.
7. Kelloff, G. J.; Boone, C. W.; Crowell, J. A.; et al. *J. Cell. Biochem.* **1996**, 26S, 1-28.
8. Huang, M. T.; Smart, R. C.; Wong, C. Q.; Conney, A. H. *Cancer Res.* **1988**, 48, 5941-5946.
9. Steele, V. E.; Moon, K. C.; Lubert, R. A. et al. *J. Cell. Biochem.* **1994**, 20, 32-54.
10. Srimal, R. C.; Dhawan, B. N. *J. Pharm. Pharmacol.* **1973**, 25, 447-452.
11. Sharma, O. P. *Biochem. Pharmacol.* **1976**, 25, 1811-1812.
12. Srivastava, R. *Agents Actions* **1985**, 28, 298-303.
13. Reddy, S.; Aggarwal, B. B. *FEBS Lett.* **1994**, 341, 19-22.
14. Korutla, L.; Kumar, R. *Biochim. Biophys. Acta.* **1994**, 1224, 597-600.
15. Huang, T. S.; Lee, S. C.; Lin, J. K. *Proc. Natl. Acad. Sci. USA.* **1991**, 88, 5292-5296.
16. Liu, J. Y.; Lin, S. J.; Lin, J. K. *Carcinogenesis* **1993**, 14, 857-861.
17. Korutla, L.; Cheung, J. Y.; Mendelsoln, J.; Kumar, R. *Carcinogenesis* **1995**, 16, 1741-17453.
18. Jiang, M. C.; Yang-Yen, H. F.; Yen, J. J. Y.; Lin, J. K. *Nutr. Cancer* **1996**, 26, 111-120.
19. Kuo. M. L.; Huang, T. S.; Lin, J. K. *Biochim. Biophys. Acta.* **1996**, 1317, 95-100.
20. Jiang, M. C.; Yang-Yen, H. F.; Lin, J. K. *Oncogene* **1996**, 13, 609-616.
21. Chen, Y. C.; Kuo, T. C.; Lin-Shiau, S. Y.; Lin. J. K. *Mol. Carcinogenesis* **1996**, 17, 224-234.
22. Cheng. S.; Wang, Z.; Ho, C. T. In : Current Medicine in China Beijing : The People's Medical Publishing House. **1988**, pp 165-172.
23. Lee,S. F.; Liang, Y. C.; Lin, J. K. *Chem. Biol. Interact.* **1995**, 98, 283-301.
24. Katiya, S. K.; Aqarwal, R.; Mukhtar, H. *Cancer Lett.* **1994**, 79, 61-66.
25. Rushmore, T. H.; Morton, M. R.; Pickett, C. B. *J. Biol. Chem.* **1991**, 266, 11632-11639.
26. Prestera, T.; Talalay, P. *Proc. Natl. Acad. Sci. USA.* **1995**, 92, 8965-8969.
27. Yu, R.; Jiao, J. J.; Duh, J. L. et al. *Carcinogenesis* **1997**, 18, 451-456.
28. Lin, Y. L.; Juan, I. M.; Chen, Y. L.; Liang, Y. C.; Lin, J. K. *J. Aqric. Food Chem.* **1996**, 44, 1387-1394.
29. Carpenter,G. *Annu. Rev Biochem.* **1987**, 56, 881-914.
30. Gill, G. N.; Bertics, P. J.; Santon, J. B. *Mol. Cell Endocrinol.* **1987**, 51, 169-186.
31. Liang, Y. C.; Lin-Shiau, S. Y.; Chen, C. F.; Lin, J. K. *J. Cell. Biochem.* **1997**, 66, in press.
32. Lin, Y. L.; Lin, J. K. *Mol. Pharmacol.* **1997**, 52, in press.
33. Burdon, R. H. *Biochem. Soc. Transact.* **1996**, 24, 1028-1032.

INDEXES

Author Index

241

Subject Index

Bestsellers from ACS Books

The ACS Style Guide: A Manual for Authors and Editors (2nd Edition)
Edited by Janet S. Dodd
470 pp; clothbound ISBN 0–8412–3461–2; paperback ISBN 0–8412–3462–0

Writing the Laboratory Notebook
By Howard M. Kanare
145 pp; clothbound ISBN 0–8412–0906–5; paperback ISBN 0–8412–0933–2

Career Transitions for Chemists
By Dorothy P. Rodmann, Donald D. Bly, Frederick H. Owens, and Anne-Claire Anderson
240 pp; clothbound ISBN 0–8412–3052–8; paperback ISBN 0–8412–3038–2

Chemical Activities (student and teacher editions)
By Christie L. Borgford and Lee R. Summerlin
330 pp; spiralbound ISBN 0–8412–1417–4; teacher edition, ISBN 0–8412–1416–6

Chemical Demonstrations: A Sourcebook for Teachers, Volumes 1 and 2, Second Edition
Volume 1 by Lee R. Summerlin and James L. Ealy, Jr.
198 pp; spiralbound ISBN 0–8412–1481–6
Volume 2 by Lee R. Summerlin, Christie L. Borgford, and Julie B. Ealy
234 pp; spiralbound ISBN 0–8412–1535–9

The Internet: A Guide for Chemists
Edited by Steven M. Bachrach
360 pp; clothbound ISBN 0–8412–3223–7; paperback ISBN 0–8412–3224–5

Laboratory Waste Management: A Guidebook
ACS Task Force on Laboratory Waste Management
250 pp; clothbound ISBN 0–8412–2735–7; paperback ISBN 0–8412–2849–3

Reagent Chemicals, Eighth Edition
700 pp; clothbound ISBN 0–8412–2502–8

Good Laboratory Practice Standards: Applications for Field and Laboratory Studies
Edited by Willa Y. Garner, Maureen S. Barge, and James P. Ussary
571 pp; clothbound ISBN 0–8412–2192–8

For further information contact:
Order Department
Oxford University Press
2001 Evans Road
Cary, NC 27513
Phone: 1-800-445-9714 or 919-677-0977
Fax: 919-677-1303